Base-level Impact

Dan Bowman

Base-level Impact

A Geomorphic Approach

Springer

Dan Bowman
Ben-Gurion University of the Negev
Beer-Sheva, Israel

ISBN 978-3-031-24993-8 ISBN 978-3-031-24994-5 (eBook)
https://doi.org/10.1007/978-3-031-24994-5

Cover illustration: Aerial view of a sequence of well preserved, wave-cut shore terraces, evidence of the dramatic base-level fall of the Upper-Pleistocene terminal Lake Lisan, 70–15 ka BP (Photo Ofek-air, Netanya, Israel)

This Springer imprint is published by the registered company Springer Nature Switzerland AG
The registered company address is: Gewerbestrasse 11, 6330 Cham, Switzerland

This book is devoted to my parents
Mrs. Trude bowman,
Dr. Robert Schneider and Dr. Ernst Bowman
and to Ruth,
Michal, Ehud and Tomer

Preface

During the emergence of an orogen, concomitant counteracting denudation processes start sculpturing the morphology. Landscape depends as much on destructive forces of erosion and weathering as on the constructive power of tectonic uplift. The final shaping of the globe is accomplished by incision of river networks fundamentally linked to base-levels that function as a principal determinant in controlling relief. Lastly, entrenchment activated by base-level reshapes the globe, forming the local relief out of mega tectonic orogens. Base-levels have thus a primary influence on landscape evolution.

The response of landscapes to tectonic motions and to base-level fluctuations is mediated, amongst others, by a "migrating morphology" and understanding the transmission signals is crucial (Jerolmack and Paola 2010). In the populated world, adjustment to a falling base-level by entrenchment may threaten property and become a devastating event, damaging agricultural lands, undermining bridges, exposing buried pipeline crossings, including water-supply facilities and destroying roads.

The aim of this volume, which is a chapter in the *fluviomorphology* topic, is to focus on the relation between base-level and the drainage basin it dominates and thereby present the base-level impact, mainly in the continental interiors. Marine base-level fluctuations along continental margins and sequence stratigraphic tract models in Exxon sequence stratigraphic sense (Koss 1994; Posamentier et al. 1988; Posamentier and Vail 1988; Posamentier 1999; Wood et al. 1993; Van Wagoner et al. 1990) are beyond the scope of this book.

This volume attempts to brings into a single integrated manuscript, principles and conclusions scattered in the literature and refer to base-level, gained by fieldwork or laboratory studies. The goal is to improve our understanding the fundamentals of the channel adjustment to the base-level control by transient signals communicated upstream through the drainage network. The chapters include presentation of different types of base-levels, discussions on the constraints of the altitude of base-levels, on responses of degradation and aggradation, on temporal and spatial trends of channel adjustment and on the knickpoint transient retreat processes and rates. **A few short, primer chapters, that focus on unique aspects, are included**. The regional approach is provided focused on the Dead Sea Area following its extreme base-level conditions which make it a unique field laboratory and an example of a recent extreme base-level fall initiated by a short history of human disturbances.

The text is relevant to students in Earth Sciences as well as to hydrologists and engineers planning and maintaining the infrastructure of roads, water pipes and power lines across incising channels and those dealing with the surface drainage. The volume will aid in environmental planning along shores of retreating lakes and in dam removal (Doyle et al. 2003).

Beer-Sheva, Israel Dan Bowman

References

Doyle MW, Stanley EH, Harbor JM (2003) Channel adjustments following two dam removals in Wisconsin. Water Resources Research 39(1)

Jerolmack DJ, Paola C (2010) Shredding of environmental signals by sediment transport. Geophysical Research Letters 37

Koss JE, Ethridge FG, Schumm SA (1994) An experimental study of the effects of base-level change on fluvial, coastal plain and shelf systems. J. sediment. Res. B64: 90–98

Posamentier HW (1999) Fundamental concepts of sequence stratigraphy

Posamentier HW, Vail PR (1988) Eustatic controls on clastic deposition II—sequence and systems tract models

Posamentier HW, Jervey MT, Vail PR (1988) Eustatic controls on clastic deposition I—conceptual framework

Van Wagoner JC, Mitcham RM, Campion KM, Rahmanian D (1990) Siliciclastic sequence stratigraphy in well logs, cores, and outcrops: concepts for high-resolution correlation of time and facies

Wood LJ, Ethridge FC, Schumm SA (1993) The effects of rate of base-level fluctuation on coastal plain, shelf, and slope depositional systems: an experimental approach. In: Sequence Stratigraphy and Facies Associations. Spec Publ. Int. Assoc. Sedim. 18

Acknowledgments

Thanks goes to **Noam Greenbaum** and to **Dalia Arad** who reviewed the manuscript, for their provocative, professional and intellectual input. This book gained as well from the experience and intelligence of **Roni Blaustein**, Cartographer and Graphic Editor for preparing the figures. Logistic assistance was granted by **Rachel Zimmerman**, the Departmental Administrator. **Oron Guy**, the System Administrator, was very helpful in providing his expertise. All are deeply thanked for an excellent cooperation. The funds for preparing the manuscript were granted by Ben-Gurion University of the Negev, Beer-Sheva, Israel.

Contents

About the Author

Dan Bowman is Professor Emeritus at the Ben-Gurion University of the Negev, Beer-Sheva, Israel. He got his Ph.D. in geomorphology at the Hebrew University. His publications include morphology of sandy coasts, nearshore circulation, tectonic morphology and alluvial fans. His main areas of research included the Israeli Mediterranean coast, the Dead Sea rift, Italian coasts and the Tien-Shan ridge in Kyrgyzstan, central Asia. Professor Bowman taught courses in nearshore circulation, coastal morphology, alluvial fans, tectonic morphology and system approach. In 2019, Springer published his book *Principles of Alluvial Fan Morphology.*

Base-Level—Definitions and Location

1

Abstract

Base-level was approximated as the sea level where the fluvial processes cease. Last Glacial Maxima (LGM) marks the last global marine low glacial base-level, −120 to −130 m below the present sea level. The rising global sea level to its recent level realized at 18–6 ka. Lakes or playas in the continental interiors compose a different base-level category. Resistant outcrops or waterfalls are local base-levels which have always an ultimate base-level downstream. The base-level functions as an external forcing on the drainage system and controls its erosive energy.

Keywords

Ultimate base-level • Local base-level • Regional base-level • Equilibrium surface • Accommodation space • Endorheic system

Base-level is a fundamental concept and a key geomorphological factor, introduced by Powell (1875). He meant the global sea level which is the grand final sink and the lowest base-level possible on earth, toward which the subaerial erosion proceeds. The concept of base-level was often, although not only, used in the context of marine settings. Base-level was approximated as sea level, and the concept *base-level change* became equivalent to the concept *sea level change* (Posamentier et al. 1988). Subsequently, the site of the base-level became the shoreline, where the fluvial processes cease and the hydraulic gradient is reduced to zero (Posamentier and Allen 1999; Catuneanu 2006). Although most streams enter the sea with considerable shear velocity and are able to erode their channel locally below, the sea serves as a downstream level boundary, which is a *control point* (Howard and Kerby 1983).

The *marine base-level* category is hardly stable at any given locality and time but is constantly changing and shifting across the continental margins and controlling major parts of the continents. The last extreme state of the global marine sea-level includes the penultimate last interglacial maximum sea level high stand,

D. Bowman, *Base-level Impact*,
https://doi.org/10.1007/978-3-031-24994-5_1

Marine Isotope Stage (MIS) 5e, 115–130 ka BP, when global sea level reached elevations of +3 to +6 m. The Last Glacial Maxima (LGM) at about 25–18 ka BP (MIS 2) marks the last global marine low base-level glacial stand 120–130 m below the present sea level. The last deglaciation with a rapidly rising global base-level to its recent level realized 18–6 ka, after which global sea level has remained very close to its present level (Lambeck and Chappell 2001).

The **second** base-level category includes larger rivers into which tributary streams flow, enclosed depressions, lakes or playas, completely detached from global sea level changes and from shore-reaching rivers. These intracontinental fluvial/lacustrine environments are defined as *endorheic* (internally drained) basins and are of an individual, free base-level, independent of the sea level and either higher or lower than it. Endorheic basins can be empty, without any filling, water basins or filled by sediment accumulation which leads to an uplift of the sedimentary surface. The transition from empty or water basins to sedimentary basins can lead as well to subsidence of the floor and further promote development of boundary faults, thus influencing the regional tectonics (Yu and Guo 2019).

Modern life added a man-made, **third** base-level category, as dammed lakes. Oceans are defined as *ultimate base-levels*.

When there is no lower area for a drainage network to drain to, the definition *ultimate* holds, including inland. The ultimate base-level is the down-channel barrier below which the river cannot cut. Every sink can be regarded as *ultimate* if it is final for the system it drains. These include internally drained dry *basins, lakes* or *swamps.* Each ultimate base-level has an *accommodation space* or *volume*, which enables sediment accumulation, marked by the base-level area (Jervey 1988). Located below the depth of incision, sediments deposited in the accommodation space have the highest potential of preservation (Blum and Törnqvist 2000). Changes in base-level area result in changes in the accommodation space and govern the geometry of the deposited sediment body. Hinterland catchments, located on footwalls, nourish their accommodation space located on the subsiding hanging wall.

Base-level was often understood as an imaginary, slightly *inclined plane*, extending below the land as the ultimate lowest level toward which denudation and river processes tend theoretically to lower and reduce earth (Posamentier and Vail 1988). Above this inclined plane particles cannot come to rest, and below it, no further wearing down is possible but only deposition and burial (Sloss 1962). Such imaginary surface, of the lowest slope along which rivers are still able to flow and transport their sediment load, was also defined as an *equilibrium surface* (Davis 1902; Wheeler 1964).

Within drainage basins, a relative resistant outcrop of bedrock, a landslide concentration of boulders (Fig. 1.1) or a reservoir lake beyond a dam form a *local base-level.* In karstic environments, when gradation of cave passages can be confirmed to a nearby river, it functions as a local base-level. Deposited sediments in caves can indicate the altitude of the controlling river base-level (Granger et al. 2001).

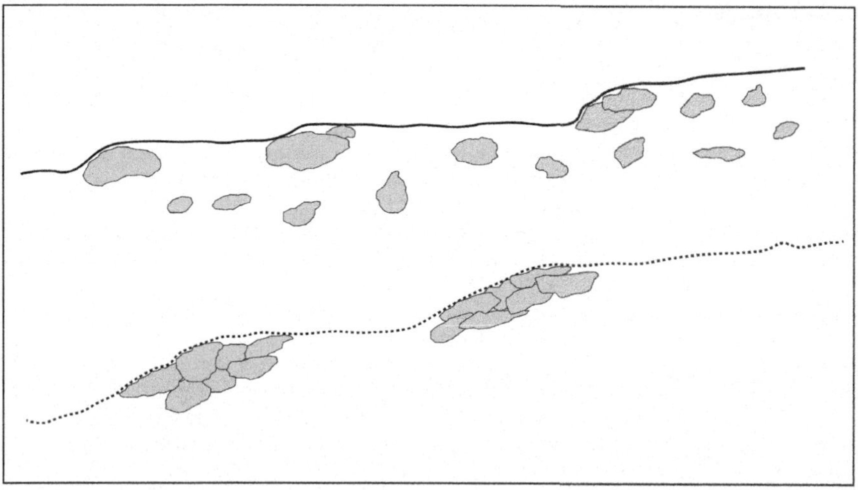

Fig. 1.1 Possible influence of concentrations of scattered boulders on the formation of mini, local base-levels in a form of small cascades during low flow along an incising channel. Each cascade controls, as a local base-level, its upslope reach (Schematic, inspired by Whipple et al. 2000)

A local base-level is to be found between the headwaters and the ultimate base-level downstream. Local base-levels are crossed by the water and sediment flow on their way downstream. Local base-levels serve as a boundary that separates the upper drainage network from the downcurrent ultimate base-level control. The upstream network is completely controlled by the highest local base-level and is independent of the ultimate downstream base-level impact. Hence, local base-levels often become the *primary control* of drainage networks.

Hanging valleys in former glaciated terrains teach us that the master valley does not yet control the tributaries as a local base-level. Anthropogenic or natural dams and road-cuts across slopes function as local base-levels too. Lateral migration of a trunk river toward its tributary results in profile foreshortening, steepening and triggering incision, i.e., a sort of a local base-level lowering effect (Harvey 2002).

A *regional base-level* relates to a larger scale. It regulates a whole region and may include wide areas with several drainage networks. All base-levels, regional, ultimate and local, force an *allogenic control*, i.e., an external impact on the drainage network controlling its erosive energy (Schumm et al. 1984, 1993; Posamentier and Allen 1999; Catuneanu 2006). Distinguished from autogenic mechanisms, the allogenic triggering will be more regional and synchronous (Daniels 2008). The base-level can be defined a *downstream control*, i.e., regulating the drainage system from the lower reaches upwards. *Upstream factors* include upper drainage area properties and climatic-hydrologic events (Blum and Törnqvist 2000) that control the drainage basin from the upper reaches downstream (Mather et al. 2017).

References

Blum MD, Törnqvist TE (2000) Fluvial responses to climate and sea-level change: a review and look forward. Sedimentology 47:2–48

Catuneanu O (2006) Principles of sequence stratigraphy. Elsevier, Amsterdam

Daniels JM (2008) Distinguishing allogenic from autogenic causes of bed elevation change in late Quaternary alluvial stratigraphic records. Geomorphology 101(1–2):159–171

Davis WM (1902) Base-level, grade and peneplain. J Geol 10(1):77–111

Granger DE, Fabel D, Palmer AN (2001) Pliocene−Pleistocene incision of the Green River, Kentucky, determined from radioactive decay of cosmogenic 26Al and 10Be in Mammoth Cave sediments. Geol Soc Am Bull 113(7):825–836

Harvey AM (2002) The role of base-level change in the dissection of alluvial fans: case studies from southeast Spain and Nevada. Geomorphology 45(1–2):67–87

Howard AD, Kerby G (1983) Channel changes in Badlands. Geol Soc Am Bull 94(6):739–752

Jervey MT (1988) Quantitative geological modeling of siliciclastic rock sequences and their seismic expression. In: Wilgus CK et al (eds) Sea level changes—an integrated approach. Society of economic paleontologists and mineralogists, Special Publication, vol 42, pp 47–69

Lambeck K, Chappell J (2001) Sea-level change through the last glacial cycle. Science 292:679–686

Mather AE, Stokes M, Whitfield E (2017) River terraces and alluvial fans: the case for an integrated quaternary fluvial archive. Quatern Sci Rev 166:74–90

Posamentier HW, Allen GP (1999) Siliciclastic sequence stratigraphy: concepts and applications. In: SEPM concepts in sedimentology and paleontology no 7, 210p

Posamentier HW, Vail PR (1988) Eustatic controls on clastic deposition II-sequence and systems tract models

Powell JW (1875) Exploration of the Colorado River of the West and its tributaries

Schumm SA (1993) River response to base-level change: implications for sequence stratigraphy. J Geol 101(2):279–294

Schumm SA, Harvey MD, Watson CC (1984) Incised channels, morphology, dynamics and control. Water Resources Publications, 200p

Sloss LL (1962) Stratigraphic models in exploration. J Sediment Res 32(3):415–422

Wheeler HE (1964) Base-level, lithosphere surface, and time-stratigraphy. Geol Soc Am Bull 75:599–610

Whipple KX, Hancock GS, Anderson RS (2000) River incision into bedrock: mechanics and relative efficacy of plucking, abrasion, and cavitation. Geol Soc Am Bull 112(3):490–503

Yu X, Guo Z (2019) The role of base level, watershed attribute and sediment accumulation in the landscape and tectonic evolution of the Circum-Tibetan Plateau basin and orogen system. J Asian Earth Sci 186

Constraints of the Base-Level Altitude

2

Abstract

The movement or stability of a base-level is always defined relative to the area drained to it, and base-levels are commonly not stable in their relative altitude. The externally imposed controlling factors are primarily tectonics, climatic-driven eustatic changes and man-made controls. Increase in stream power, following base-level fall, is a principal result that triggers an upstream signal of adjustment in form of a geomorphologic impact. Tectonic uplifts should be viewed as base-level falls and vice versa. Uplifts may cause altitude contrasts between flanking base-levels of the same orogen. Subsidence of the supplying drainage system means a relative rise of the base-level. The area draining toward a base-level may be stable, rising fast or slow relative to the base-level, or subsiding (Vail et al.1977). Mountain relief is finally controlled by the interplay between rising or falling relief. Eustatic fluctuations of lake levels will cause an increase or decrease in the accommodation space and are usually more frequent than tectonically driven base-level fluctuations.

Keywords

Tectonic base-level change • Inland lakes • Culverts • Man-made control • Artificial cutoff • Accommodation space • Eustatic movement

Base-levels are dynamic surfaces, commonly not stable in their altitude for very long periods, controlled by allogenic and internal non-steady factors. Base-level change, i.e., the vertical altitudinal change, may depend on (1) climatic-hydrologic-sedimentologic systems, all external and (2) on movements of structural elements, i.e., on tectonics which represents an internal driven system. The imposed controlling factors include as well man-made controls. Hence, base-level changes may be *extrinsic controlled or intrinsic controlled* by tectonics, lithology or slope gradient. Water level curves of lakes following changes in evaporation, precipitation and runoff are *eustatic* changes. To the vertical water plane changes we should add the

D. Bowman, *Base-level Impact*,
https://doi.org/10.1007/978-3-031-24994-5_2

vertical earth movements, including differential isostatic rebound following loading and unloading (Baedke and Thompson 2000). On the continents across the globe we receive as a result large base-level height changes induced by different combinations of eustatic, isostatic, volcanic and tectonic contributions (Slangen et al. 2012). Base-level changes are not globally uniform because the spatial variability in the different contributions is very large. Even in the sea, local level change may deviate substantially from the global mean.

Change in the base-level height is a change in the boundary conditions to which the entire drainage network will have to adjust. The base-level has a primary stabilizing or changing impact on the drainage network. For example, formation of a *peneplain*—a flat, exhumed planation surface—requires and indicates base-level stability (Phillips 2002). The altitude changes of base-levels discussed in this volume are always *relative* changes, relative to the nearby earth's surface, i.e., to the surrounded drained areas. Slowly sinking basins surrounded by rapid sinking blocks will form a rising base-level impact. Increase in stream flow gradient and in the stream power are the principle responses to base-level fall (Bull 1991). Base-level change, followed by steepening and power increase, will trigger an upstream signal of adjustment (Crosby 2006).

2.1 Tectonic/Isostatic and Volcanic Base-Level Control

Tectonics and base-level are not independent from each other. Fluvial systems are sensitive to isostatics and tectonically imposed boundary conditions which change the *relative altitude* of the base-level. During a tectonic phase, uplift may exceed the destructional erosion. Although material is transported away from the highlands toward the base-level the absolute elevations may increase. The tectonic uplift and subsidence are translated into geomorphological base-level impacts. A tectonic uplift should be viewed as a base-level fall and vice versa. Volcanic activity is relevant too. Subsequent to emplacement of lava flows, new higher surfaces and drainage networks develop, controlled by a now relative lower base-level.

A tectonic base-level fall can result just from uplift along a bounding fault (Fig. 2.1). Uplift of a footwall relative to a downthrown hanging wall means base-level fall (Fig. 2.2). Uplifted highlands and nearby subsiding sedimentary basins will evolve with abrupt channel deepening and steepening. Isostatic uplift in response to erosional unloading of the crust and nearby concomitant depositional loading and subsidence will increase the macro-relief and the effect of base-level fall (Bishop and Brown 1992). Low depth of incision toward base-level in mountains will decrease their isostatic rebound.

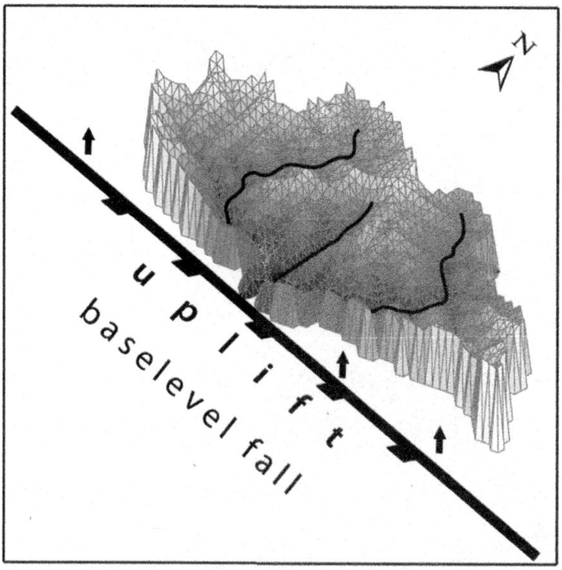

Fig. 2.1 Uplifted channels on a footwall along a bounding fault demonstrate a lowered tectonic base-level (modified after Attal et al. 2008)

Base-level uplift causes a decrease in the channel gradient and rivers tend to abandon their channel and change their route. Following uplift, rivers may change their course away from areas of the greatest localized uplift into different basins and base-levels. A rising base-level will occur following relief subsidence. Isostatic adjustments to changes in crustal loading are slow and may result in long-term uplifts or subsidences. Rivers that run parallel to structures typically avoid crossing faults or changes in lithology. Rivers responding to increased uplift fault slip rates develop diagnostic morphological responses similar to those activated by a falling base-level.

The morphology of an emerging landscape will finally be the result of the combined effects of the uplift pattern and the erosional response. Base-level fall or rise during an orogenic uplift history is a principal control of the final morphology initially dictated by the fault geometry. Tectonic uplifts may cause altitude differences between flanking base-levels of an orogen, and the relief may become divided unevenly between the flanks because of the base-level altitudinal contrasts (Fig. 2.3). Drainage divide positions will shift too, following the erosional competition between flanks with different mean gradients. The altitude of the base-level and the different lengths of the drainage basins, i.e., the different distances of the base-level from the divide, which constrain the topographic declivity, will trigger erosion and modify the tectonic topography differently, leading to an asymmetric topography that no longer reflects the tectonic structure. The mountain relief will finally be controlled by the interplay between fault geometry and the different base-levels control. Hence, both fault geometry and the base-levels exert important controls on present morphology (Barnes et al. 2011; Densmore et al. 2007).

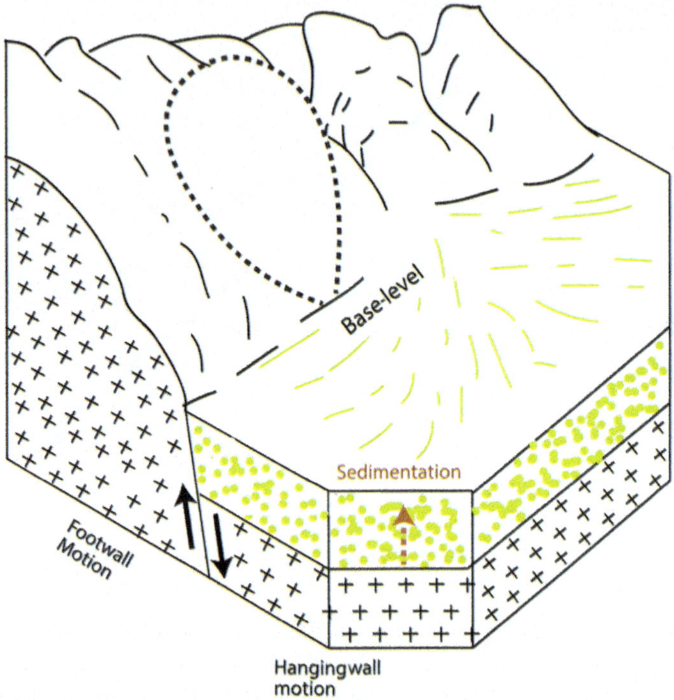

Fig. 2.2 A drainage basin located on an uplifted footwall with its base-level on the subsiding hanging wall. The subsidence rate of the base-level is set by the rate of the footwall uplift relative to the rate of the hanging wall subsidence minus the rate of the sedimentation (redrawn from Goren et al. 2014)

2.2 Climatic Base-Level Control

Climatic changes such as glacial and interglacial periods triggered lowering and rising base-levels, respectively. The recent interglacial stage caused a significant rise of the marine base-level and valleys flooding. In the continental environments climatic changes induced base-level changes through fluctuations in lake levels and lake desiccation. Multiple eustatic lake level fluctuations can be ascribed to fluctuations induced by the waxing and waning of the Pleistocene glaciers. During the glacial recession, as long as the outflow from lakes exceeded the inflow from rivers, i.e., a negative water balance, lakes' level constituted an eustatic lowering base-level and decreased in the accommodation space. The climatic changes may vary on much shorter timescales than tectonic uplifts and thus are usually more frequent than tectonically driven base-level fluctuations.

Variations in the levels of inland lakes are governed by interactions between the shape of the catchment, the amount of water and sediments entering from the drainage basin and the amount of available accommodation space. The geomorphology of the lake area determines how changes in climate-related water input

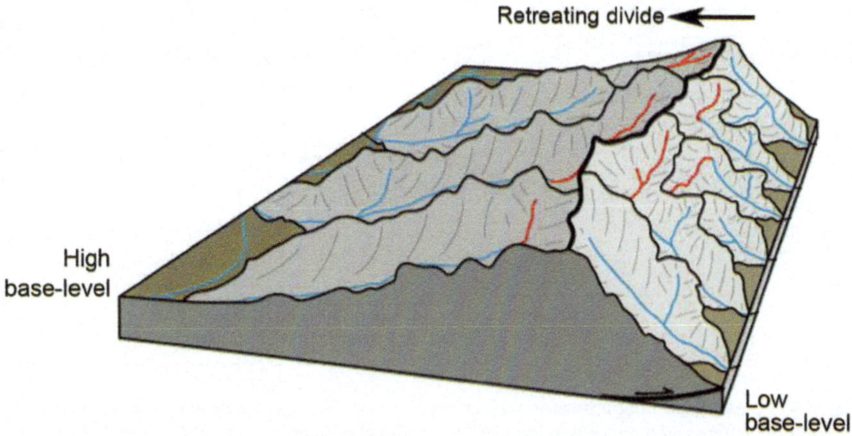

Fig. 2.3 Asymmetric tectonic relief shaped along its flanks by base-levels of different altitudes or distance (modified after Seagren and Schoenbohm 2019). The morphological contrasts between the mountain flanks are demonstrated by different catchment gradients and drainage spacing (Burbank and Pinter 1999). The divide will be shifted away from the lower and more effective base-level that controls the steepest flank. Different drainage reorganization on the flanks is demonstrated by different widening or narrowing of the drainage basins and different outlet spacings. Given a steeper flank, outlet spacing will decrease through the splitting of the drainage basins (Bonnet 2009). Apparent regularity of the drainage pattern can be seen especially on the steep side of the mountain. The morphology and the site of the divide are an interplay between tectonics and the base-level-controlled processes (Ellis and Densmore 2006)

will change the lake level. Minor climatic changes can produce large changes in the base-level altitude due to a relatively small lake. Accumulation in playas—internal drainage basins—constitutes a base-level rise which may bury the piedmont fronts, including fronts of alluvial fans. Vertical base-level change can be extreme: lake level changes of 300 m in 10–20 ky were common throughout the Pleistocene.

2.3 Man-Made Control

Dam construction, reservoir drawdown and gravel mining along channels are man-made base-level changes. Managed base-level lowering includes culverts under highways when crossing floodplains or active fan areas. An excavated culvert below the elevation of the surrounding surface forms a local new base-level that triggers headward incision (Fig. 2.4). To minimize their geomorphic impact, culverts should not be excavated in highway management practices. Highways should be raised above the elevation of the active surface so that culverts under the highway will be placed at their natural grade (Florsheim et al. 2001).

Rivers that are straightened and steepened by construction of artificial cutoffs respond as if a local lower base-level was formed. The decrease in length through the cutoffs results in a noticeable increase in the local gradient, i.e., >1

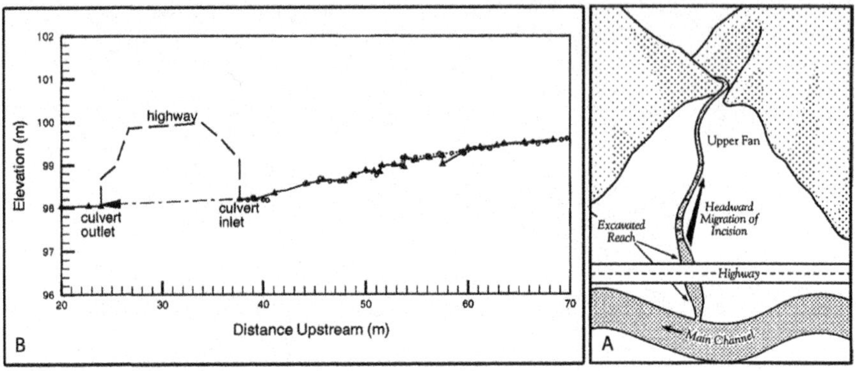

Fig. 2.4 To avoid forming a local low base-level that would trigger headward incision, channels should not be excavated under the highway (A). Flow should proceed freely on the surface below heightened highways (B) (from Florsheim et al. 2001)

times greater than the bend way (Winkley 1977; Hooke 2004; Turnipseed 2017). The response in the Lower Mississippi River to such artificial cutoffs showed in some cases disappearance of the reactions within 1–3 years although in other cases stability was achieved only after about 100 year (Schumm et al. 1984).

References

Attal M, Tucker GE, Whittaker AC, Cowie PA, Roberts GP (2008) Modeling fluvial incision and transient landscape evolution: influence of dynamic channel adjustment. J Geophys Res Earth Surf 113(F3)

Baedke SJ, Thompson TA (2000) A 4,700-year record of lake level and isostasy for Lake Michigan. J Great Lakes Res 26(4):416–426

Barnes JB, Densmore AL, Mukul M, Sinha R, Jain V, Tandon SK (2011) Interplay between faulting and base-level in the development of Himalayan frontal fold topography. J Geophys Res Earth Surf 116(F3)

Bishop P, Brown R (1992) Denudational isostatic rebound of intraplate highlands: the Lachlan river valley, Australia. Earth Surf Proc Land 17(4):345–360

Bonnet S (2009) Shrinking and splitting of drainage basins in orogenic landscapes from the migration of the main drainage divide. Nat Geosci 2(11):766–777

Bull WB (1991) Geomorphic responses to climatic change. Oxford University Press, Oxford, p 326

Burbank DW, Pinter N (1999) Landscape evolution: the interactions of tectonics and surface processes. Basin Res 11(1):1–6

Crosby BT (2006) The transient response of bedrock river networks to sudden base-level Fall. PhD thesis submitted to the Massachusetts Institute of Technology

Densmore AL, Allen PA, Simpson G (2007) Development and response of a coupled catchment fan system under changing tectonic and climatic forcing. J Geophys Res Earth Surf 112(F1)

Ellis MA, Densmore AL (2006) First-order topography over blind thrusts. Special papers, Geological Society of America, 398, 251 p

Florsheim JL, Mount JF, Rutten LT (2001) Effect of base-level change on floodplain and fan sediment storage and ephemeral tributary channel morphology, Navarro River, California. Earth Surf Process Land J Br Geomorphol Res Group 26(2):219–232

Goren L, Fox M, Willett SD (2014) Tectonics from fluvial topography using formal linear inversion: theory and applications to the Inyo Mountains, California. J Geophys Res Earth Surf 119(8):1651–1681

Hooke JM (2004) Cutoffs galore! Occurrence and causes of multiple cutoffs on a meandering river. Geomorphology 61(3–4):225–238

Phillips JD (2002) Erosion, isostatic response, and the missing peneplains. Geomorphology 5(3–4):225–241

Schumm SA, Harvey MD, Watson CC (1984) Incised channels: morphology, dynamics and control. Water Resources Publications

Seagren EG, Schoenbohm LM (2019) Base-level and lithologic control of drainage reorganization in the Sierra de las Planchadas, NW Argentina. J Geophys Res Earth Surf 124(6):1516–1539

Slangen ABA, Katsman CA, Van de Wal RSW, Vermeersen LLA, Riva REM (2012) Towards regional projections of twenty-first century sea-level change based on IPCC SRES scenarios. Clim Dyn 38(5):1191–1209

Turnipseed C (2017) Characterizing the hydrodynamics of a meandering river neck cutoff. Master thesis, Louisiana State University

Vail PR, Mitchum RM Jr, Thompson S III (1977) Seismic stratigraphy and global changes of sea level: part 4. Global cycles of relative changes of sea level. Section 2. Application of seismic reflection configuration to stratigraphic interpretation

Winkley BR (1977) Man-made cutoffs on the lower Mississippi river, conception, construction, and river response, no. 300-2. Army Engineer District Vicksburg, Corps of Engineers

3 Discontinuity in Slope—Knickpoints and Knickzones

Abstract

Knickpoint is referred to a slope break in a channel or a change in bed elevation. Knickpoints develop when crossing a hard lithology, a tectonic line, an injection site of coarse sediments or at tributary junctions. Waterfalls are vertical knickpoints that have a tread, a lip, a face and a downstream reach, often with a mid-stream bar and a plunge pool. The drawdown space is the oversteepened reach of the longitudinal profile above the knickpoint face. Knickzone is an oversteepened channel reach, often including rapids or pools, that is typically up to a few kms long. Nickzones, as knickpoints, are often transient features, eroding and migrating upstream with time as a response to a base-level fall.

Keywords

Knickpoint/Knickzone frequency and density • Knickpoint partitioning • Waterfall • Tread • Face • Lip • Plunge pool • Drawdown space

3.1 Knickpoints

According to Walther Penck's lexicon (Von Engeln 1940), *Knickpunkte* are abrupt changes in the longitudinal profile of a channel detected through high slope values that make breaks in the overall profile. Knickpoints are the locally steep sections of the longitudinal profile, such as waterfalls, that are the sites of the greatest concentration of energy dissipation. A common combination of bed slopes accompanying a knickpoint is that of a gentle reach upstream from the knickpoint and a rather steep slope downstream (Fig. 3.1). The knickpoint indicates a maximum shear value. Downstream from this point the water depth decreases and the slope increases. The erosion potential is maximal directly over the knickpoint (Brush and Wolman 1960). A software called *Knickpoint Finder* that uses a digital elevation model (DEM) and provides a fundamental advantage in automatic identification of knickpoints along drainage profiles had been suggested (Queiroz et al. 2015).

D. Bowman, *Base-level Impact*,
https://doi.org/10.1007/978-3-031-24994-5_3

The ratio between the oversteepened slope below the knickpoint to the average channel slope is an index of the relative magnitude of the slope disturbance. Knickpoints form a reach with a convex profile but may as well indicate only a break in *bed elevation* (Haviv et al. 2010). *Frequency* and *density* of knickpoints are their number and length percentage along a unit river length, respectively. Experiments suggest that knickpoint frequency may be an indicator of the base-level fall rate (Grimaud et al. 2016). A thick shielding of the bedrock by alluvium and a slow base-level fall rate will lengthen periods without knickpoint formation. Knickpoints are considered as non-equilibrium markers. Lack of knickpoints suggests a more advanced graded state (Baldwin et al. 2003; Bishop et al. 2005; Mackey et al. 2014)

The origin of knickpoints has been already discussed by Penck (1925), and it was further debated whether knickpunkte will develop and persist while migrating headward (Von Engeln 1942). The Knickpoints are characterized by the following multiple causality (Phillip et al. 2010):

Lithologic contrasts—(Figs. 3.1 and 3.2a) Very subtle variations in erosion resistance can cause the development of a knickpoint. It cannot be overemphasized that lithologic heterogeneities are a critical factor in the initiation of knickpoints. Zones of high knickpoint density are often located in areas rich in lithological contacts. Knickpoints often form by the harder more resistant layer overlying a softer more erodible one. The lithologic or structural change make a relative *static knickpoint* (Harbor et al. 2005). Knickpoints often highlight lithological contacts as their main control without any base-level impact (Whipple 2004). However,

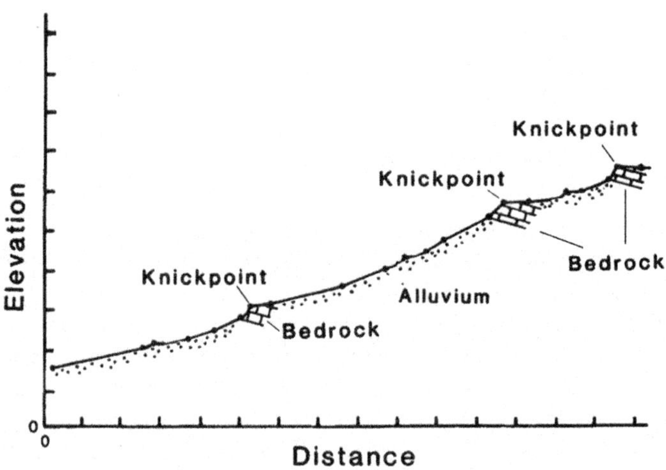

Fig. 3.1 A longitudinal channel bed profile composed of alternating alluvium and hard lithology which form lithologic-controlled knickpoints. The knickpoints are typically located between a gentle reach upstream and a steeper downstream. The ratio of the typical oversteepened slope below the knickpoint to the average slope of the profile is an index of the relative magnitude of the disturbance (Redrawn from Miller et al. 1991)

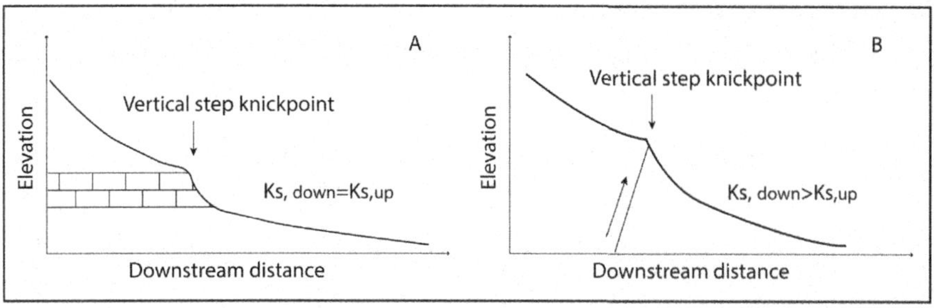

Fig. 3.2 Different channel steepness (Ks) may form upstream and downstream of knickpoints. When crossing a more resistant lithology, a knickpoint is expected **(A)** shown by a convexity with similar up and down steepness. Following a tectonic uplift **(B)** a knickpoint forms with higher steepness below (Redrawn from Lague 2014)

when it comes to knickpoint dynamics, the substrate factors may turn out to be of secondary importance compared to the catchment area factor.

Tectonic activity—This includes spatial variations in uplift rate (Fig. 3.2b), crossing geological structures and channel orientation changes relative to the dip (Gardner1983; Miller 1991). A strong linearity of the pattern of knickpoints distribution may suggest association with faulting. The tectonic activity may cause only a break in the channel bed *elevation* without changes in slope (Haviv et al. 2010).

Transient wave of erosion ("Erosional/Fluvial knickpoint")—(Fig. 3.3) knickpoints may be the front of an upstream-propagating incision, i.e., a transient feature that transmits the response to a base-level fall in a form of an erosion steepening wave that moves from the channel outlet through the drainage basin toward the headwaters. This had been recognized already in the early years of the twentieth century (Davis 1933; Penck 1925; Baulig 1940).

Injection of coarse sediments—Due to frequent debris flows or large slope failures, including large-scale landsliding (Lamb et al. 2007; Schumm et al. 1984).

Tributary junctions—The gap between channels of different incision capabilities is bridged by a knickpoint located at the junction between the tributary and its deeper incised trunk stream. Knickpoints also form at capture sites.

The knickpoint group falls generally into two categories: (1) convexities in the channel profile that are anchored in space and tend to evolve in place rather than migrate upstream. These knickpoints include sites of change in the substrate erodibility, at landslide dams, by debris flows, at tributary junctions or when crossing tectonic lines. (2) The second category includes migrating knickpoints that transient (migrating upward) through the network in response to perturbations such as base-level fall (Morell et al. 2012).

Identification of knickpoints and examination their potential relationship with controlling factors can be accomplished by using longitudinal profiles plotted from digital elevation model (DEM) data, by analyzing landslide hazard maps including

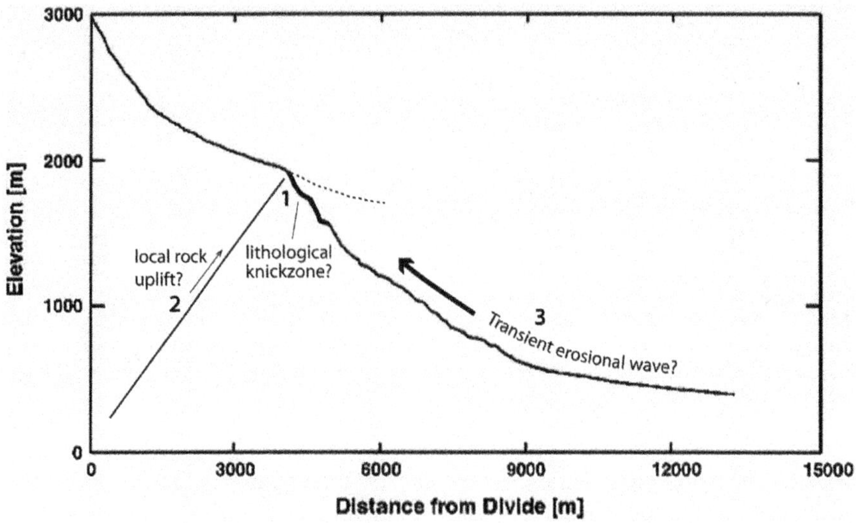

Fig. 3.3 A knickpoint in the longitudinal channel profile may be shaped by either: **1**. a more resistant lithology—a lithological knickzone, or: **2**. an increase in the local rock uplift rate. The lack of an obvious fault or lithologic contact may suggest that the knickpoint may be: **3**. a transient erosional signal propagating upstream as a response to base-level fall (Modified after Tucker and Whipple 2002; Lague 2014)

active and inactive ones and colluvial slopes susceptible to mass movements, by studying soil surveys maps, geologic maps with lithological boundaries and structural features as well as investigating aerial photographs and field observations (Phillips et al. 2010).

Vertical knickpoints (waterfalls) and their associated plunge pools are frequent features. The energy gained as the water drops over these knickpoints is dissipated in plunge pools. The mechanism of scour below waterfalls depends on the balance between the substrate resistance and the imposed hydraulic stresses. The effect of bedload on the scour of resistant, non-alluvial beds shows by generation of sculpted bedforms such as large channels and bedrock benches, longitudinal grooves, potholes and flutes. Clear-water scour does erosive work below overfalls by direct impact of the free jet of plunging water. Even highly resistant concrete slabs are erodible by the direct impact of a hydraulic jet, with the rate of scour diminishing as the angle of attack deviates from perpendicular (U.S. Bureau of Reclamation 1948; Pasternack et al. 2007). When a knickpoint is quasi-vertical (a waterfall) water and sediment fall do not apply direct force on the knickpoint face. In that case, waterfall recession is influenced by a variety of other processes, including plunge pool drilling, freeze–thaw, wet-dry cycles and groundwater seepage (see Chap. 7).

Waterfalls are composed of (Fig. 3.4): (1) an upstream bedrock reach - the *tread* - characterized by a bedrock surface where the overlying alluvial materials have been removed. It ends upstream where the channel floor starts to be buried by alluvium. (2) The mid-reach vertical *face* that extends from the *lip* that delimits its upper end to the base of the knickpoint and may be composed of a few strata. (3) The *downstream reach* includes the plunge pool that often merges with a midstream bar where the coarse debris is usually deposited (Miller 1991). We may expect undercutting and vertical headwall failures due to the hydraulic stresses in the plunge pool region (Wells et al. 2009); however, knickpoints must not necessarily be undercut with a plunge pool.

The *Drawdown space* is the oversteepened reach of the longitudinal profile above the knickpoint where the water surface starts to steepen toward the lip which is the upper spike of the break in slope. According to Haviv et al. (2006), upslope of the lip the flow is accelerating toward a freefall as a result of the low pressures at the lip cross section. The oversteepened reach is thus due to enhanced incision

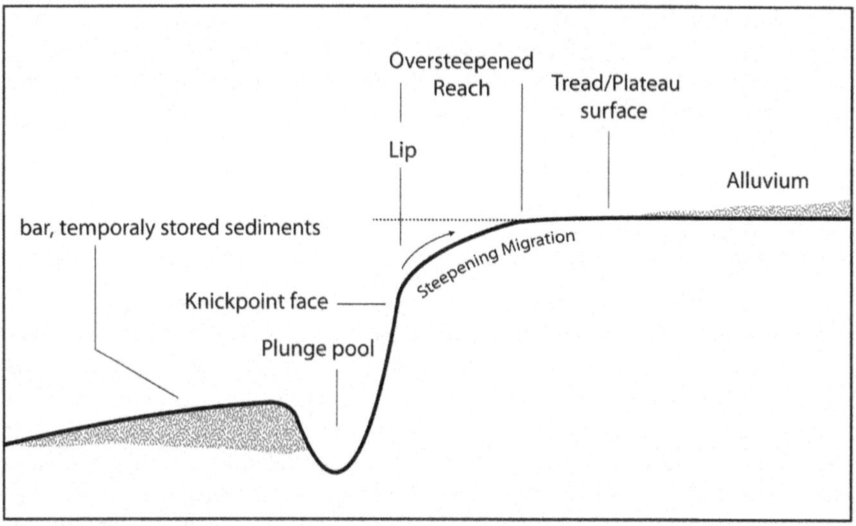

Fig. 3.4 Definition sketch of waterfall components with a plunge pool (schematic). The draw-down space is the oversteepened reach upslope of the lip where flow accelerates and thins toward the knickpoint freefall. Lowering the lip is lowering the base-level for the upstream plateau. The oversteepened reach grows upstream, thus increasing the channel slope relative to the former plateau surface (dashed line). Steepening upstream indicates the incision signal moving upstream, i.e., the information about downstream fluvial incision is leaking upstream through the knickpoint. Length of the oversteepened segments is usually a few hundred meters. It is about 1300 m long above the Niagara. Over short length scales (tens to hundreds of meters) steepening result from the flow acceleration at the free overfall. Over longer length scales (kilometers) structural and lithologic controls may also contribute to steepening (Haviv et al. 2006; Berlin et al. 2009; Redrawn from Begin 1979)

related to the drawdown effect. The flow over the knickpoint reduces the convexity while increasing the slope of the channel upstream. As a result, the knickpoint will degrade and incision will propagate upstream over time at ever decreasing rate. Such changes could also reflect rock strength properties. The hydraulic drawdown stretches a few hundred meters to a few kms up-lip and corresponds to the transition from the alluvial to the bedrock reach (Weissel and Seidl 1998). Similar steepening was observed in flume laboratory experiments in non-cohesive material by Gardner (1983) and Brush and Wolman (1960).

Approaching the knickpoint lip from upstream, the channel width decreases while flow velocity and the bottom shear stress increase, reaching a maximum and exceeding the critical shear strength of the bedrock and thus allowing the entrainment of the unconsolidated channel bed alluvium from upstream of the lip that becomes an exposed bedrock reach (Miller 1991). The lip may be linear when it occurs along a single joint but may as well become nonlinear when the lip follows the path of vertical conjugate joints. Further upstream, above the waterfall lip, the longitudinal profile evolution has been studied by numerical simulation (Haviv et al. 2006).

In micro-morphology scale, headcuts of rills form micro-knickpoints, i.e., small steps in the bed surface. Gullies end as well upslope by knickpoint headcuts. Coastal scarps, where shoreline abuts cliffs, may compose waterfall knickpoints when attacked by waves from below and crossed by an incising channel from above.

3.2 Knickzones

Besides a knickpoint, the base-level fall signal may also be transmitted upstream as a *knickzone* which is a relative *long oversteepened channel reach* that often exhibit channel narrowing and may develop over hundreds of meters up to several hundreds of kilometers. Such reach can be composed of rapids, cascades and plunge pools or pools and riffles. Its bed roughness on which sediment transport is strongly dependent, shows strong positive correlations with the bed slope. Knickzones are characterized as well by changes in bed cover grain size, in channel geometry and in erosion processes such as plunge-pool drilling, toppling and undercutting (Figs. 3.5 and 3.6).

Knickzones may be composed of multiple faces and treads that together resemble a stair sequence, dependent on the bedrock stratigraphic thickness. Knickzones as steeper channel segments have greater energy expenditure than have the nearby above and below reaches (Foster 2010) and mark a segment where the stream gradients are steeper than expected for the contributing drainage area. When a knickzone lengthens as it propagated upstream, its local slope decreases through time. *Knickzone density* defines the percentage of the knickzone reach length out of a total given stream length (Hayakawa and Oguchi 2014).

Knickzones are more expected along steep river reaches underlain by harder rocks. An increase in knickzone gradient is expected with increasing altitudes

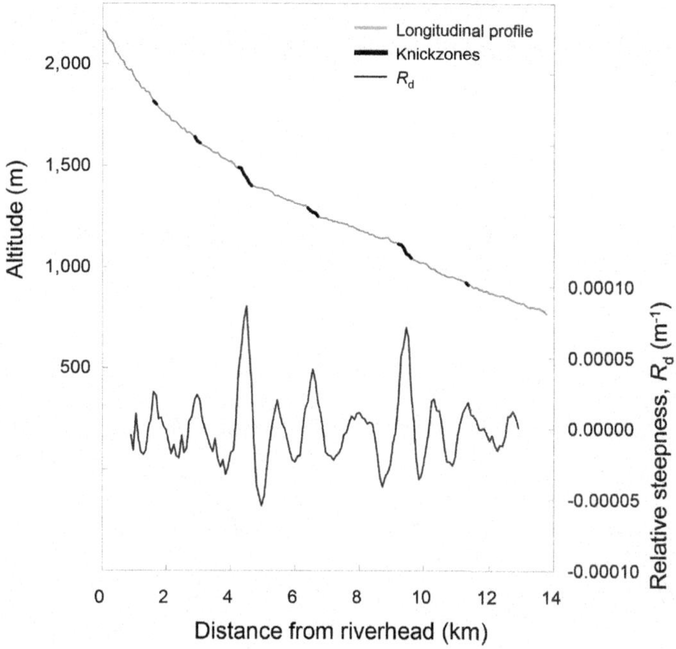

Fig. 3.5 Longitudinal profile with identified knickzones along the Matsukawa River, central Japan. Segments of less than 1 m height are not shown. Knickzones are shown as local steep segments between the gentle segments above and below. Steepness was defined relative to the horizontal length used for the gradient calculation (From Hayakawa and Oguchi 2006)

(Hayakawa and Oguchi 2006). Watersheds close to active faults comprise more knickzones. In piedmont areas knickzones are often consistent with tectonic faulting at the mountain foot. Lithology may too exert influence on the knickzone abundance.

The identification of knickzones requires individual selection of the steepened reaches, i.e., locate the lips and the corresponding bases of the knickzone along the longitudinal stream profile. An algorithm has been constructed to automate this procedure by objectively selecting the bounds, map the position and measure the knickzone dimensions (Neely et al. 2017). An objective criterion to determine knickzones' upper and lower limits was also suggested by Hayakawa and Oguchi (2006, 2009) based on analysis of DEMs and GIS and on a threshold value of a relative steepness.

Upstream drainage networks respond more slowly to a base-level fall because the erosional signal is dampened by knickzones and knickpoints. When propagating headwards throughout the drainage network, knickzones become shorter and steeper and when passing a tributary outlet often do not propagate through the tributary. When knickpoints are breached, especially where the channel substrate is most resistant, they turn into a deep, narrow channel with high gradients and

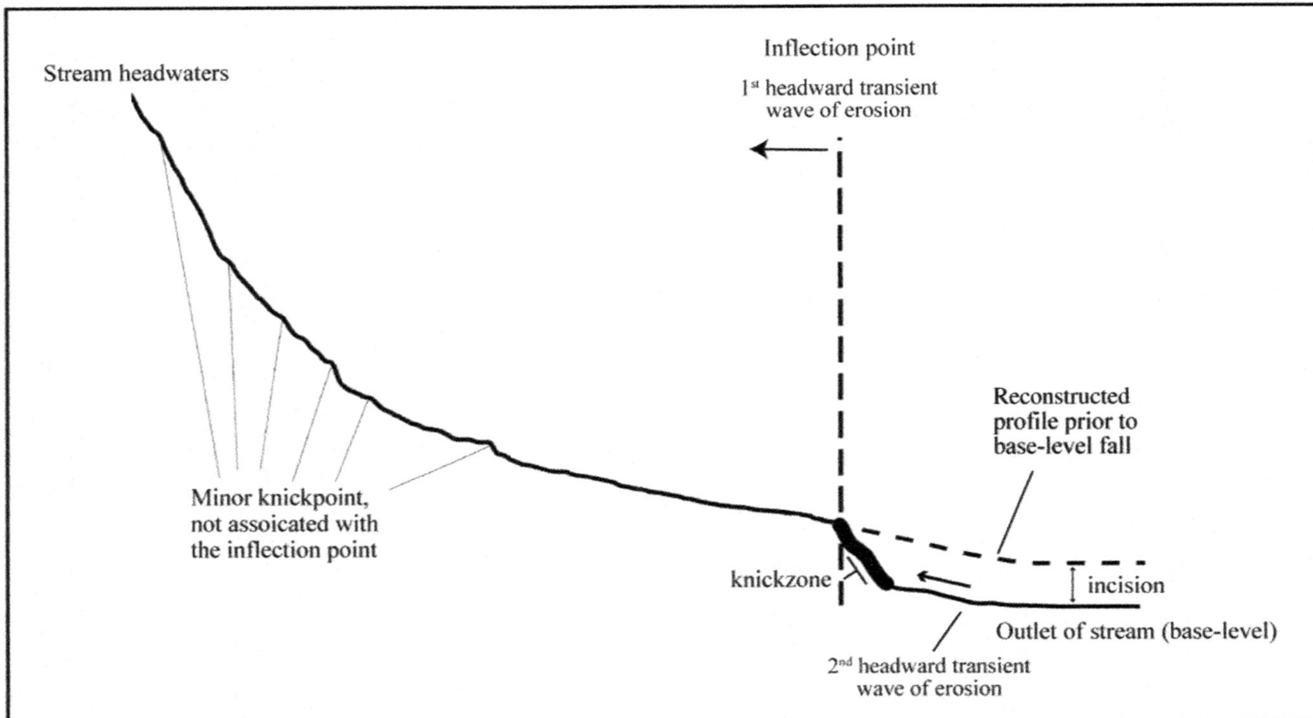

Fig. 3.6 Schematic figure depicting a knickzone and minor knickpoints upstream following two stages of evolution (Modified from Foster and Kelsey 2012). Base-level fall and incision can be estimated by projecting the stream profile downstream of the inflection point and subtracting the modern active profile elevation, using concavity similar to the regional average and steepness indices that are locally representative

stream power, i.e., become a knickzone in form of a narrow, actively incising gorge-like valley (Wohl et al. 1999).

References

Baldwin JA, Whipple KX, Tucker GE (2003) Implications of the shear stress river incision model for the timescale of post orogenic decay of topography. J Geophys Res 108(B3)
Baulig H (1940) Reconstruction of stream profiles. Geomorphology 3:3–15
Begin ZB (1979) Aspects of degradation of alluvial streams in response to base-level lowering. Unpublished Ph.D. dissertation, Colorado State University Fort Collins, Col. 239p
Bishop P, Hoey TB, Jansen DJ, Artza IL (2005) Knickpoint recession rate and catchment area: the case of uplifted rivers in Eastern Scotland. Earth Surf Processes Landforms 30(6):767–778
Brush LM, Gordon Wolman M (1960) Knickpoint behavior in noncohesive material: a laboratory study. Geol Soc Am Bull 71(1):59–74
Davis WM (1933) Piedmont benchlands and primaerruempfe. Geol Soc Am Bull 43:399–440
Foster M (2010) Knickpoints in tributaries of the South Fork Eel River, Northern California, Doctoral dissertation, Humboldt State University
Foster M, Kelsey HM (2012) Knickpoint and knickzone formation and propagation, South Fork Eel River, Northern California. Geosphere 8(2):403–416
Gardner TW (1983) Experimental study of knickpoint and longitudinal profile evolution in cohesive, homogeneous material. Geol Soc Am Bull 94(5):664–672
Grimaud JL, Paola C, Voller V (2016) Experimental migration of knickpoints: influence of style of base-level fall and bed lithology. Earth Surf Dyn 4(1):11–23
Harbor D, Bacastow A, Heath A, Rogers J (2005) Capturing variable knickpoint retreat in the Central Appalachians, USA. Geogr Fis Din Quat 28(1):23–33
Haviv I, Enzel Y, Whipple KX, Zilberman E, Stone J, Matmon A, Fifield LK (2006) Amplified erosion above waterfalls and oversteepened bedrock reaches. J Geophys Res: Earth Surf 111(F4)
Haviv I, Enzel Y, Whipple KX, Zilberman E, Matmon A, Stone J, Fifield KL (2010) Evolution of vertical knickpoints (waterfalls) with resistant caprock: insights from numerical modeling. J Geophys Res: Earth Surf 115(F3)
Hayakawa YS, Oguchi T (2006) DEM-based identification of fluvial knickzones and its application to Japanese mountain rivers. Geomorphology 78(1–2):90–106
Hayakawa YS, Oguchi T (2009) GIS analysis of fluvial knickzone distribution in Japanese mountain watersheds. Geomorphology 111(1–2):27–37
Hayakawa YS, Oguchi T (2014) Spatial correspondence of knickzones and stream confluences along bedrock rivers in Japan: implications for hydraulic formation of knickzones. Geogr Ann Ser B 96(1):9–19
Lague D (2014) The stream power river incision model: evidence, theory and beyond. Earth Surf Proc Land 39(1):38–61
Lamb MP, Howard AD, Dietrich WE, Perron JT (2007) Formation of amphitheater-headed valleys by waterfall erosion after large-scale slumping on Hawai. Geol Soc Am Bull 119(7–8):805–822
Mackey BH, Scheingross JS, Lamb MP, Farley KA (2014) Knickpoint formation, rapid propagation and landscape response following coastal cliff retreat at the last interglacial sea-level highstand: Kaua'I, Hawai. Geol Soc Am Bull 126(7–8):925–942
Miller JR (1991) The influence of bedrock geology on knickpoint development and channel-bed degradation along down cutting streams in south-central Indiana. J Geol 99(4):591–605
Morell KD, Kirby E, Fisher DM, Van Soest M (2012) Geomorphic and exhumational response of the Central American Volcanic Arc to Cocos Ridge subduction. J Geophys Res: Solid Earth 117(B4)

Neely AB, Bookhagen B, Burbank DW (2017) An automated knickzone selection algorithm (KZ-Picker) to analyze transient landscapes: calibration and validation. J Geophys Res Earth Surf 122(6):1236–1261
Pasternack GB, Ellis CR, Marr JD (2007) Jet and hydraulic jump near-bed stresses below a horseshoe waterfall. Water Resources Res 43(7)
Penck WO (1925) Die piedmontflächen des südlichen schwarzwaldes: Gesellschaft für Erdkunde, Berlin V.I:81–108
Phillips JD, McCormack S, Duan J, Russo JP, Schumacher AM, Tripathi GN, Pulugurtha S (2010) Origin and interpretation of knickpoints in the Big South Fork River basin, Kentucky–Tennessee. Geomorphology 114(3):188–198
Queiroz GL, Salamuni E, Nascimento ER (2015) Knickpoint finder: a software tool that improves neotectonic analysis. Comput Geosci 76:80–87
Schumm SA, Harvey MD, Watson CC (1984) Incised channels, morphology, dynamics and control. Water Resources Publications, 200p
Tucker GE, Whipple KX (2002) Topographic outcomes predicted by stream erosion models: sensitivity analysis and intermodel comparison. J Geophys Res: Solid Earth 107(B9):ETG-1
U.S. Bureau of Reclamation (1948) Model studies of spillways, Bull 1Boulder Canyon Project, final reports, part VI, Hydraulic investigations Denver, Colo
Von Engeln OD (1940) A particular case of knickpunkte. Ann Assoc Am Geogr 30(4):268–271
Von Engeln OD (1942) Geomorphology. MacMillan and Company, New York, 655p
Weissel JK, Seidl MA (1998) Inland propagation of erosional escarpments and river profile evolution across the southeast Australian passive continental margin. In: Tinkler KJ, Wohl EE (eds) Rivers over rock: fluvial processes in bedrock channels. Geophysical Monograph 107, American Geophysical Union, pp 189–206
Wells RR, Alonso CV, Bennett SJ (2009) Morphodynamics of headcut development and soil erosion in upland concentrated flows. Soil Sci Soc Am J 73(2):521–530
Whipple KX (2004) Bedrock rivers and the geomorphology of active orogens. Annu Rev Earth Planet Sci, 32:151–185.
Wohl EE, Thompson DM, Miller AJ (1999) Canyons with undulating walls. Geol Soc Am Bull 111(7):949–959

Degradation

4

Abstract

For a channel to become graded to a lowered base-level it must incise. After base-level drop, incision and transport capabilities increase and will gradually be transmitted upstream. The balance between sediment supply and sediment transport efficiency determines the response of the channel to base-level perturbations. Bedrock erosion is proportional to the stream energy available and depends on the channel slope and discharge and inversely proportional to river width. The rate of incision sets the rate for hillslope processes as well. The sequence—base-level fall, entrenchment, sediment migration and armor coverage—are expected to be repeated by cycles of base-level fall. An accelerating base-level fall restricts lateral channel mobility and forms incised narrow valleys and gorges. While narrowing and steepening are the key response for increasing the erosive effect of rivers flowing over harder lithology adapting to base-level fall, lateral channel incision is expected toward more stable base-level conditions.

Keywords

Vertical wearing • Gorge • Armor ratio • Lateral incision • Stream power • Autogenic/Allogenic processes

4.1 Incision

The main scientific interest in base-level is related to the trend of triggering incision by its fall. The efficiency of river incision is complex, being related to climate, relief, flow hydraulics, sediment transport, discharge events and lithology as well as to weathering. The level of cohesion and the grain size of the sediments are also directly connected to erodibility. There are several ways in which a river may respond, at least partly, to small base-level lowering, aside incision. These include change of: the cross-sectional channel width, the bed roughness, the extent of the alluvial cover, the bed material grain size, the degree of the sinuosity and the bed

D. Bowman, *Base-level Impact*,
https://doi.org/10.1007/978-3-031-24994-5_4

morphology (Whipple 2004; Gardner 1973; Schumm 1993, 2005; Blum 1993). If base-level fall is large, the ultimate result is incision and triggering a *degradation wave* to migrate in the upstream direction.

Incision had been experimentally demonstrated through lowering the base-level at flumes' outlet (Begin 1978; Begin et al. 1980) and can be well observed in gullies (Leopold and Miller 1956). It should, however, be emphasized that incision along channels is not controlled only by a lowering base-level. Different factors, unconditioned by base-level, such as changes in water discharge or decrease in the sediment supply may as well trigger degradation. Decline in sediment load when, for example, sediment becomes strapped, will enhance incision significantly.

It had been concluded, based on field and experimental studies, that when base-level in front of a stream outlet is lowered and the slope gradient is steep, the elevation potential provides the stream the energy to erode backward. *Threshold shear stress* is the minimum bed shear stress required to initiate detachment of bed material. The channel will incises in order to reduce its steep gradient and become graded to the lowered base-level (Leopold and Bull 1979). The amount of incision depends on the amplitude of the base-level drop (Marriott 1999). The effect of a falling base-level that exposes a steep slope was initially studied in experiments that demonstrated processes across continental margins, once the shoreline dropped below the shelf break. The lowering of the base-level along steeper gradients accelerates flow velocities and incision. Bedrock incision may only require a minor increase in slope beyond which is necessary to transport the supplied coarse load (Sklar and Dietrich 2001).

When the gradient of the new exposed segment following base-level fall is not different from the profile upstream, the dominating aggradational or degradational trends will not change. If, following base-level fall, the exposed gradients are insufficient to trigger incision, sediments may simply prograde and bury the newly exposed frontal surface. The event will be defined as a *non-dissecting base-level fall*. Headward incision across exposed areas, following base-level fall, triggered development of dendritic drainage systems (Koss et al. 1994; Bowman 2010; See Chap. 16, Fig. 16.7).

The balance between sediment supply and sediment transport efficiency determines the response of a channel to perturbations. Streams may incise if their transport capacity—a function of discharge and slope—will be greater than the sediment load supplied from upstream. An increased sediment supply may limit the extent of the exposed bedrock by partial burial (Ethridge et al. 2005) but may as well provide tools for abrasion of the exposed bedrock. Bedload sediment flux in non-alluvial channels may often be less than the capacity load. Channel beds may be alluvium floored by sand, coarse gravel, alternating bedrock and alluvium, or be only bedrock (Howard 1998; Howard et al. 1994) which is often locally exposed at knickpoints as the resistant oversteepened reach (Whipple and Tucker 2002).

The rate of incision determines the rate of base-level lowering for hillslope processes which deliver their debris into the channel. Alluvial and bedrock sections may alternate along channels seasonally or during longer terms. Channels are often characterized by a coherent blanket of transportable sediment cover over the bedrock and before bedrock incision can be initiated, the release of the alluvial cover should be accomplished. So long as erosion amount is less than the thickness of the alluvium, the channel remains alluvial. When the alluvial channel cover is stripped, a bedrock channel will form.

Incision processes pose hazards to man-made structures, especially to bridges by vertical deepening and widening of channels. Fluvial erosion of bedrock may occur via *corrosion*, which refers to chemical weathering and by *solution. Abrasive weathering* is gained by sediment as bedload or suspended load is transported along the channel. Physical experiments have shown that small particles as silt and sand traveling in suspension cause less erosion than larger particles moving as bedload (Sklar and Dietrich 2001). The notion that erosion rates by suspension are orders of magnitude lower than erosion rates by equivalent amounts of gravel and larger particles has been propagated into numerical models (Sklar and Dietrich 2004; Lamb et al. 2008).

Bedrock incision occurs primarily due to abrasion by saltating and rolling bedload. The most efficient abrasive gravel size is the intermediate grain size, large enough to travel but not too large to be immobile (Sklar and Dietrich 2001). *Cavitation* occurs when velocity fluctuations in a flow induce pressure changes that weaken the bedrock and pit the rock surface. Cavitation is enhanced along joints, bedding plane or other surface irregularities in bedrock. *Turbulence* and lift forces contribute to the quarrying of jointed rocks. Most models of river erosion are based on the premise that the rate of stream incision can be expressed as a function of stream gradient S, the drainage area A and some additional factor such as the sediment load. The energy per time available to erode is defined as the *stream power* (Leopold et al. 1964; Bull 1979) proportional to the local channel slope as well as to the local drainage area.

Winnowing and selective transportation of the finer particles, when the coarse grains are left relatively immobile, form a surface layer *armor* (Dietrich et al. 1989), i.e., an erosion-resistant reach of coarse material that limits channel incision and temporally prohibits knickpoint migration. Such surface coarsening is typical of river reaches that are degrading due to a deficit in sediment supply. The expansion of coarse patches, resulting in coarsening of the streambed, is usually examined by the ratio between surface and subsurface grain-size metrics (the *armor ratio*). Selective transport that coarsens the bed is expected during the recession limbs of long-lasting floods with declining sediment supply. The sequences—base-level fall, entrenchment, sediment propagation and armor coverage—are expected to be repeated by cycles of base-level falls (Embry 2009; Shanley and MacCabe 1991). When armoring is removed, the exposed sediments are more easily transported, and the sediment load increases.

Erosion threshold as a critical stream power or shear stress must be exceeded for erosion to occur. The incision rate varies with either stream power per unit bed area (Hancock et al. 1998; Whipple and Tucker 1999) or with stream power per unit channel length (Seidl and Dietrich 1992). Bedrock erosion is proportional to the stream energy available to erode by water moving downstream, dependent on the channel slope and on the river discharge and inversely proportional to river width, i.e., $rgQS/W$, where r = density of the water, g = gravitational acceleration, Q = water discharge, S = channel slope commonly approximated as water surface slope and W = channel width. Entrained material may be limited by the ability of the stream to detach bed material, or by its capacity to transport which is a function of discharge and slope.

Degradation occurs when stream power is more than sufficient to transport the bedload, whereas aggradation will start when stream power is insufficient. The stream power incision rule assumes that transport capacity well exceeds the imposed sediment load and the rate of channel incision is limited simply by the capacity of detaching bedrock fragments from the channel bed (Howard and Kerby 1983; Whipple and Tucker 1999). General models are based on the hypothesis that the incision rate should be proportional to either the total stream power, or to the basal shear stress (Howard et al. 1994; Whipple and Tucker 1999). Transport-limited erosion assumes that the rate of surface lowering is limited by the rate at which sediment particles can be transported away.

When channel incision occurs, hillslope steepen and debris are detached and pulled down, delivered into the channel and become bedload that facilitates abrasion processes. Vertical incision includes the removal of fractured fragments of bedrock that can be plucked out by fluid shear stress, without the aid of sediment impact (Chatanantavet and Parker 2009; Sklar and Dietrich 2004). Parker (2004) presented an incision model in which two processes dominate erosion of the channel bed: plucking of bedrock blocks and abrasion by saltating bedload. Hydraulic wedging of small clasts into cracks contributes as well to the loosening and removal of joint blocks (Whipple et al. 2000). When incision by abrasion and plucking, assisted by joint spacing fractures and bedding planes, is very slow, bedrock weathering may become a competing process.

4.2 Deepening and Narrowing

Hydraulic geometry is sensitive to changes in the boundary conditions and can serve as a tool to understand the response of channels to base-level fluctuations. In slow base-level fall conditions, lateral migration plays a larger role leading to wider channels. Channel widening characterizes decelerating base-level fall and may continue to widen during base-level rise. An accelerating base-level fall restricts lateral channel mobility, incising narrow valleys or canyons with minor lateral migration (Strong and Paola 2008). Field observations (Snyder and Kammer 2008; Lamb and Fonstad 2010) and experimental studies of channel incision (Bigi et al. 2006; Johnson and Whipple 2007) showed that knickpoints coincide with channel narrowing and steepening. As the substrate resistance increases, erosion becomes more localized in the form of potholes or longitudinal grooves and the bed gradient becomes steeper.

Channel narrowing across a zone of active uplift (i.e., a tectonic base-level fall), was found through areas of rapid uplift as the central Himalayas (Lavé and Avouac

2000). Field surveys in eastern Tibet support the hypothesis that the transition to deep gorges is systematically related to departure from the classical channel width versus drainage area relationship (Montgomery 2004). The transition from wide alluvial to narrow bedrock channels follows a change to slope dependency. Gorges and canyons, with typical narrow and deep cross sections with steep gradients are most common downstream of knickpoints. Their undulations are considered as the remnants of breached potholes or sinuous longitudinal grooves formed during incision (Wohl et al. 1999).

Channel width is thus an important mode of adjustment to base-level changes (Whipple et al. 2013) and a first-order morphological *diagnostic criteria* for an upstream migrating erosional response (Whittaker et al. 2007b). Wide channels not only bring about less focused erosion but may also increase sediment storage. In contrast, valleys with significant incision are narrow, v-shaped and without terraces (Strong and Paola 2008; Martin et al. 2009). When channels narrow, it increases the unit stream power and the shear stress and generates a higher erosive capacity. Knickpoints and knickzones, which form steep reaches, respond faster when the channel narrows up to the point where boundary friction and energy dissipation on the walls become significant and counteract any further narrowing. Channel narrowing followed by widening was demonstrated experimentally by Cantelli et al. (2004) when flow incised into reservoir deposits after removal of a dam or dam failure (Greenbaum 2007). During the initial period, under the rapid drawdown, rapid incision formed a narrow and deep channel until lateral sediment transport along the sidewalls became effective and stopped the narrowing. At this stage, the minimum width has been attained and the stream banks became unstable and failed. Beyond this time, sediment transport along the sidewalls became effective and the channel slowly widened (Fig. 4.1). A positive feedback began by deposition of point and mid-channel bars which routed flows toward the channel banks, leading bank failure, widening and decreasing vertical incision.

Valley narrowing is thus a transient (migrating upstream) response of rivers that adapt to tectonic uplift (base-level fall) by increasing entrenchment while struggling to regain stability. The overall stage development of the landscape can be understood by the degree to which channels have narrowed. Channels will narrow significantly in regions of high uplift and steepness. Where the falling base-level effect is the highest, channels are typically narrow and often with a gorge-like morphology. Channels will become narrower for the same drainage area when the rate of the uplift (base-level fall rate) is greater (Finnegan et al. 2005; Whittaker et al. 2007a; Attal et al. 2008).

Narrowing and increase in the incision rate is an intrinsic way by which channels adjust to a falling base-level. Narrowing is a positive feedback enhancing degradation. The narrow morphology can be used as an indicator of the base-level efficiency.

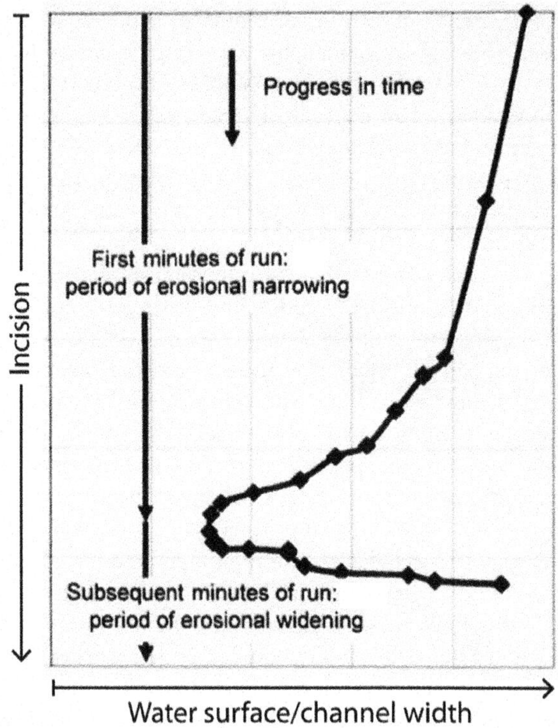

Fig. 4.1 Experimental water surface width plotted against water elevation with time demonstrating erosional narrowing followed by erosional widening. Schematic (modified from Cantelli et al. 2004)

4.3 Lateral Erosion

Based on models (Doyle et al. 2003) and on field observations (Pickup 1975), channels may widen while incising. Martin et al. (2011) noted too that the majority of valley widening in experiments occurred during base-level fall. However, at a rapid base-level drop, almost the entire stream's erosive energy is expected to be utilized in vertical cutting, giving rise to deeply incised channels with no substantial widening (Majeski 2009). Yoxall (1969), based on experimental work, suggested that the rate of lateral erosion varies directly with the rate of base-level lowering. The dominance of lateral channel incision is expected toward relative stable base-level conditions, when downcutting becomes slow enough. Slow drops in base-level tend to give rise to an initially wider and shallower channel (Schumm 1987); i.e., the main transfer of energy expenditure from vertical erosion to lateral activity is expected in the post-vertical incision events. Long-continued lateral cutting suggests that the stream is in equilibrium and that it has no excess energy for further downcutting (Strong and Paola 2008). Valleys are thus expected to be at their widest after being formed over a long period of time. During widening, a significant sediment volume is exported and may serve as a tool to erode, but can as well inhibit incision by forming a deposit cover that limits the extent of bedrock exposure along the channel bed.

When lateral erosion significantly outpaces vertical incision for an extended period of time, the channel bed may become a flat surface. Saltating bedload particles will be deflected by roughness elements on the bed and directed to the channel walls, causing bank erosion on impact. The deflection of bedload particles is the primary process driving lateral bedrock erosion (Fuller 2014). Widening of channels causes velocity decrease and limits erosion headward migration, i.e., forms a negative feedback to incision (Begin et al. 1981).

4.4 Complex Response

Adjustment of channels to base-level lowering includes incision, followed by sediment migration downstream and possible storage along the channel, making the entire response an episodic and complex event (Schumm et al. 1984). The sediment on its way downstream will be *temporally* deposited as *channel fill, the depositional response* to the erosional base-level fall effect (Martin et al. 2009). Channel fill indicates an autogenic (self-generated) complex response phenomena (Schumm 1973; Schumm et al. 1987; Parker 1977) and does not reflect allogenic forcing on the system, such as changes in eustasy, tectonics or climate.

The complex response scenario that is initiated by base-level fall and includes periods of incision followed by aggradation and terminating by renewed incision of the channel fill (Schumm et al. 1987) is an *autocyclic* process that emphasizes the complexity of the overall base-level-triggered degradation process. Valley entrenchment thus includes an ongoing interplay of incision, backfilling and channel migration (avulsion) that may continuously redefine the shape of the incised valleys. There are *internal* fluvial processes that have the same degradation effect as base-level lowering. Lateral migration of a trunk stream at a tributary junction toward the branch shortens the tributary length and has an incisional effect similar to lowering base-level. Decrease in sediment load as an internal process will increase the capability of flow entrenchment. These examples suggest that incision is possible while base-level fall is irrelevant. One should be cautious in assigning allogenic base-level controls.

References

Attal M, Tucker GE, Whittaker AC, Cowie PA, Roberts GP (2008) Modeling fluvial incision and transient landscape evolution: influence of dynamic channel adjustment. J Geophys Res Earth Surf 113:F3

Begin ZB (1978) Aspects of degradation of alluvial streams in response to base-level lowering. PhD thesis, Colorado State University, Fort Collins, 239 p

Begin ZB, Meyer DF, Schumm SA (1981) Development of longitudinal profiles of alluvial channels in response to base-level lowering. Earth Surf Proc Land 6(1):49–68

Begin ZB, Schumm SA, Meyer DF (1980) Knickpoint migration due to base-level lowering. Journal of the Waterway, Port, Coastal and Ocean Division 106(3): 369–388

Bigi A, Hasbargen LE, Montanari A, Paola C (2006) Knickpoints and hillslope failures: interactions in a steady-state experimental landscape. In: Willett SD, Hovius N, Brandon MT,

Fisher DM (eds) Tectonics, climate, and landscape evolution. USGS special paper 398, Penrose conference series. US Geological Survey, Boulder, CO, pp 295–307

Blum M (1993) Genesis and architecture of incised valley fill sequences: a late quaternary example from the Colorado River, Gulf Coastal Plain of Texas. In: Recent applications of siliciclastic sequence stratigraphy, chap 10

Bowman D (2010) The Dead Sea graben: geomorphology of the lowest spot on Earth. In: Migon P (ed) Geomorphological landscapes of the world. Springer, Berlin, pp 247–255

Bull WB (1979) Threshold of critical power in streams. Geol Soc Am Bull 90:453–464

Cantelli A, Paola C, Parker G (2004) Experiments on upstream-migrating erosional narrowing and widening of an incisional channel caused by dam removal. Water Resour Res 40(W03304):1–12

Chatanantavet P, Parker G (2009) Physically based modeling of bedrock incision by abrasion, plucking, and macroabrasion. J Geophys Res Earth Surf 114(F4)

Dietrich WE, Kirchner JW, Ikeda H, Iseya F (1989) Sediment supply and the development of the coarse surface layer in gravel-bedded rivers. Nature 340(6230):215–217

Doyle MW, Stanley EH, Harbor JM (2003) Channel adjustments following two dam removals in Wisconsin. Water Resour Res 39(1)

Embry AF (2009) Practical sequence stratigraphy. Canadian Society of Petroleum Geologists 81

Ethridge FG, Germanoski D, Schumm SA, Wood LJ (2005) The morphologica and stratigraphic effects of base-level change: a review of experimental studies. In: Blum M, Marriott S, Leclair S (eds) Fluvial sedimentology vii. Special Publication. International Association of Sedimentologists, pp 213–241

Finnegan NJ, Roe G, Montgomery DR, Hallet B (2005) Controls on the channel width of rivers: implications for modeling fluvial incision of bedrock. Geology 33(3):229–232

Fuller TK (2014) Field, experimental and numerical investigations into the mechanisms and drivers of lateral erosion in bedrock channels. Doctoral dissertation, University of Minnesota

Gardner TW (1973) A model study of river meander incision. Unpublished M.S. thesis, Colorado State University, Fort Collins, Colorado, 86 p

Greenbaum N (2007) Assessment of dam failure flood and a natural, high-magnitude flood in a hyperarid region using paleoflood hydrology, Nahal Ashalim catchment, Dead Sea, Israel. Water resources research 43(2)

Hancock GS, Anderson RS, Whipple KX, Tinkler KJ, Wohl EE (1998) Beyond power: Bedrock river incision process and form. Geophysical Monograph-American Geophysical Union 107: 35-60

Howard AD (1998) Long profile development of bedrock channels: interaction of weathering, mass wasting, bed erosion, and sediment transport. In: Geophysical monograph, vol 107. American Geophysical Union, pp 297–319

Howard AD, Dietrich WE, Seidl MA (1994) Modeling fluvial erosion on regional to continental scales. J Geophys Res Solid Earth 99(B7):13971–13986

Howard AD, Kerby G (1983) Channel changes in badlands. Geol Soc Am Bull 94(6):739–752

Johnson JP, Whipple KX (2007) Feedbacks between erosion and sediment transport in experimental bedrock channels. Earth Surf Proc Land 32:1048–1062

Koss JE, Ethridge FG, Schumm SA (1994) An experimental study of the effects of base-level change on fluvial, coastal plain and shelf systems. J Sediment Res 64(2b):90–98

Lamb MP, Dietrich WE, Sklar LS (2008) A model for fluvial bedrock incision by impacting suspended and bed load sediment. J Geophys Res 113

Lamb MP, Fonstad MA (2010) Rapid formation of a modern bedrock canyon by a single flood event. Nat Geosci 3:477–481

Lavé J, Avouac JP (2000) Active folding of fluvial terraces across the Siwaliks Hills, Himalayas of central Nepal. J Geophys Res 105:5735–5770

Leopold LB, Bull WB (1979) Base-level, aggradation, and grade. Proc Am Philos Soc 123(3):168–202

Leopold LB, Miller JP (1956) Ephemeral streams: Hydraulic factors and their relation to the drainage net. US Government Printing Office

Leopold LB, Wolman MG, Miller JP (1964) Fluvial processes in geomorphology. WH Freeman and Company, San Francisco, 507 p

Majeski AL (2009) Fluvial systems tied together through a common base-level: the geomorphic response of the Dirty Devil River, North Wash Creek, and the Colorado River to the rapid base-level drop of Lake Powell. All graduate theses and dissertations, Utah State University, Paper 291

Marriott SB (1999) The use of models in the interpretation of the effects of base-level change on alluvial architecture. In: Fluvial sedimentology VI, vol 28. Blackwell Science, pp 271–281

Martin J, Cantelli A, Paola C, Blum M, Wolinsky M (2011) Quantitative modeling of the evolution and geometry of incised valleys. J Sediment Res 81:64–79

Martin J, Paola Ch, Abreu V, Neal J, Sheets B (2009) Sequence stratigraphy of experimental strata under known conditions of differential subsidence and variable base-level. AAPG Bull 93(4):503–533

Montgomery DR (2004) Observations on the role of lithology in strath terrace formation and bedrock channel width. Am J Sci 304(5):454–476

Parker RS (1977) Experimental study of basin evolution and its hydrologic implication. Unpublished PhD dissertation, Colorado State University, Port Collins, Colorado

Parker G (2004) Somewhat less random notes on bedrock incision. Int Memo 118:20

Pickup G (1975) Downstream variations in morphology, flow conditions and sediment transport in an eroding channel. Z Geomorphol Neue Folge 19:443–459

Schumm SA (1973) Geomorphic thresholds and complex response of drainage systems. Fluvial Geomorphol 6:69–85

Schumm SA (1993) River response to base-level change: implications for sequence stratigraphy. J Geol 101(2):279–294

Schumm SA (2005) River variability and complexity, chap 16. Cambridge University Press, Cambridge, pp 149–155

Schumm SA, Harvey MD, Watson CC (1984) Incised channels: morphology, dynamics, and control. Water Resources Publications

Schumm SA, Mosley MP, Weaver W (1987) Experimental fluvial geomorphology. Wiley, New York, p 413

Seidl MA, Dietrich WE, Schmidt KH, de Ploey J (1992) The problem of channel erosion into bedrock. Functional geomorphology 23: 101–124

Shanley KW, McCabe PJ (1991) Predicting facies architecture through sequence stratigraphy-an example from the Kaiparowits Plateau. Utah Geol 19(7):742–745

Sklar LS, Dietrich WE (2001) Sediment and rock strength controls on river incision into bedrock. Geology 29:1087–1090

Sklar LS, Dietrich WE (2004) A mechanistic model for river incision into bedrock by saltating bedload. Water Resour Res 40(6)

Snyder NP, Kammer LL (2008) Dynamic adjustments in channel width in response to a forced diversion: Gower Gulch, Death Valley National Park. Calif Geol 36(2):187–190

Strong N, Paola C (2008) Valleys that never were: time surfaces versus stratigraphic surfaces. J Sediment Res 78(8):579–593

Whipple KX (2004) Bedrock rivers and the geomorphology of active orogens. Annu Rev Earth Planet Sci 32:151–185

Whipple KX, Dibiase RA, Crosby BT (2013) Bedrock rivers. In: Treatise on geomorphology. Elsevier, Amsterdam, pp 550–573

Whipple KX, Hancock GS, Anderson RS (2000) River incision into bedrock: mechanics and relative efficacy of plucking, abrasion, and cavitation. Geol Soc Am Bull 112(3):490–503

Whipple KX, Tucker GE (1999) Dynamics of the stream-power river incision model: implications for height limits of mountain ranges, landscape response timescales, and research needs. J Geophys Res Solid Earth 104(B8):17661–17674

Whipple KX, Tucker GE (2002) Implications of sediment-flux-dependent river incision models for landscape evolution. J Geophys Res Solid Earth 107(B2):ETG-3

Whittaker AC, Cowie PA, Attal M, Tucker GE, Roberts GP (2007a) Bedrock channel adjustment to tectonic forcing: implications for predicting river incision rates. Geology 35(2):103–110

Whittaker AC, Cowie PA, Attal M, Tucker GE, Roberts GP (2007b) Contrasting transient and steady-state rivers crossing active normal faults: new field observations from the Central Apennines, Italy. Basin Res 19(4):529–556

Wohl EE, Thompson DM, Miller AJ (1999) Canyons with undulating walls. Geol Soc Am Bull 111(7):949–995

Yoxall WH (1969) The relationship between falling base-level and lateral erosion in experimental streams. Geol Soc Am Bull 80(7):1379–1384

Base-Level Rise

5

Abstract

Base-level fall provides a more dramatic geomorphological effect than does a rising base-level. Relatively few studies have been concerned with the effects of base-level rise. The impact of a falling base-level extends further far up compared to the spatial limited impact of aggradation by a rising base-level. High sediment supply from rivers can restrain a river mouth from being entrenched when base-level falls and may enable, under base-level rise, continued sediment fill basinward.

Keywords

Accommodation space • Autoretreat • Intracontinental basins • Blocked-valley lakes

5.1 The Coastal Marine Environment

Following the Last Glacial Low Sea Level, the ensuing rising marine base-level brought wide areas below water level and provided a wide *accommodation space* for deposition. When marine water level (base-level) rises, accommodation increases further upstream into valleys. Blocked-valley lakes appeared to have formed in response to Pleistocene–Holocene sea level rise, forcing aggradation on the main stem sourced in the highlands. A significant marine flooding and aggradation deep into the continents may, however, not be explained only by eustatic base-level rise but may need as well an additional significant tectonic subsidence (Aubrey 1989).

Following a rising base-level, when deltas may be drowned and decoupled from the shoreline, high sediment supply from rivers can periodically overwhelm the rising base-level (Muto and Steel 2002) and deltas may continue and build forwards during a modest sea level rise. Deposition on the Mississippi submarine fan is known to have continued during Holocene rising sea level (Kolla and Perlmutter 1993). The same occurred on the Amazon Fan during the rising sea level (Flood

D. Bowman, *Base-level Impact*,
https://doi.org/10.1007/978-3-031-24994-5_5

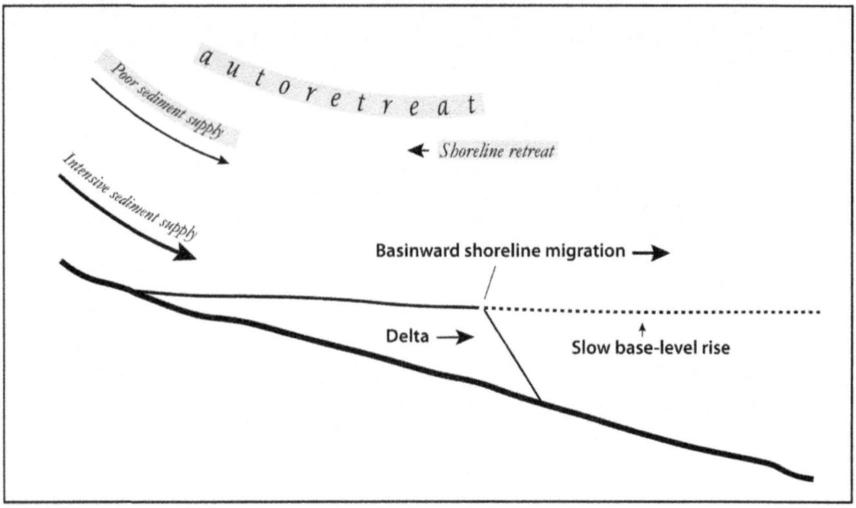

Fig. 5.1 During a rising base-level, when the rate of sediment supply to the delta is great, the shoreline migrates *basinward* and the delta is advancing. Autoretreat occurs when sediment supply is poor and following base-level rise the shoreline moves *landward*

et al. 1991). This fluvial aggradation is expected in line with a positive interplay between sediment supply and base-level rise and is expected later to stop following decrease in the overall aggradation rate because of *sediment starvation.*

Autoretreat defines shoreline migration landward following base-level rise (Fig. 5.1). It depends on the fluvial sediment supply relative to the rate and duration of a base-level rise (Parker et al. 2004). Autoretreat may occur with or without complete sediment supply starvation from the continent toward the shoreline (Heller et al. 2001). Adequate fluvial sediment supply may build up a river mouth and prevent it from been drowned due to the sea rise (Parker et al. 2004). Coastal encroachment of beach deposits may occur as a part of the landward migration of the coastal facies while onlapping (Fig. 5.2). The coastal deposits may build a few meters above the raised base-level.

5.2 Continental Aggradation

As relief is weared down, intracontinental basins start to fill-in. Continuous sediment accumulation will fill endorheic basins and raise significantly this sort of continental base-level. When lakes spread, accommodation is formed along their expanding margins. Spreading limnic facies indicates base-level rise (Vail et al. 1977; Grimaud et al. 2016). Remants of lacustrine sediments, filling canyons around the recent Dead Sea, Israel, mark the paleo-high base-level of the Upper Pleistocene Lake Lisan (see Chap. 16).

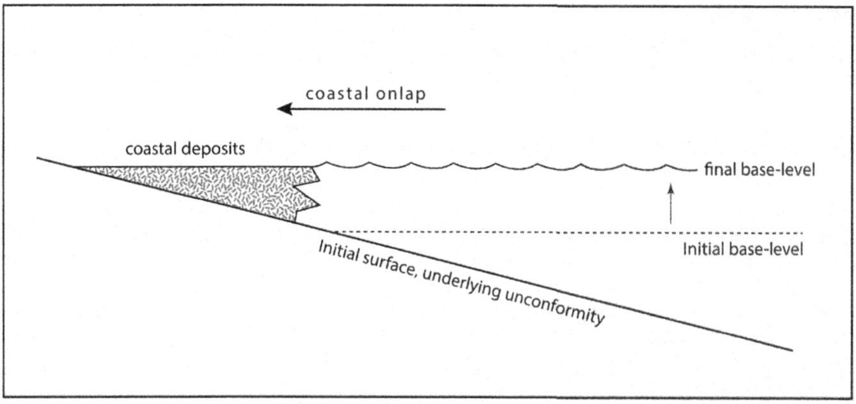

Fig. 5.2 Onlap of coastal deposits indicates the relative rise of the base-level (following Vail et al. 1977)

A water reservoir operates as a rising barrier across a stream path. The water surface trapped behind the barrier operates as a raised local base-level. The barrier traps sediments which are the local fill (Catuneanu et al. 2011). Upstream backfilling (Leopold and Bull 1979; Leopold 1992; Schumm 1993) flattens the channel bed for a short distance. The deposition extends upstream, decreases in thickness and wedges out to zero. The effect of raised base-levels can be traced only a short distance of several hundred meters upstream and not throughout the entire river network (Leopold et al. 1964); i.e., the effect does not propagate upstream beyond a limited section. Upstream of the head of the sediment wedge, the longitudinal profile is unaltered by the base-level rise. The wedged fill is enveloped between the initial profile below and the level of maximum aggradation above (Holbrook et al. 2006). Removal of dams or of log jams demonstrates the impact of a falling local base-level and conversion of the reservoir from a sediment sink to a sediment source.

A rising base-level may cause deposition downstream, and at the same time degradation will still be dominant upstream since the last glacial base-level lowering. As channels that experienced base-level fall during last glaciation may still be incising, thus their upper and lower stream reaches will be out of phase (Schumm 2005).

The impact of a rising or falling base-level was used as well to explain morphological evolutions on Mars (Cardenas et al. 2018). Evidence of fluvial erosion and deposition or incised and filled valleys on Mars was suggested to indicate falls and rises of base-level. Crosscutting valleys are explained by at least two episodes of base-level fall and rise. Records of cut and fill were suggested as response to fluctuations in a nearby body of water and increase in coarse sediments within a river system was related to base-level fall. Filling of valleys was related to coastline transgression as base-level rose.

References

Aubrey WM (1989) Mid-Cretaceous alluvial-plain incision related to eustasy, southeastern Colorado Plateau. Geol Soc Am Bull 101(4)

Cardenas BT, Mohrig D, Goudge TA (2018) Fluvial stratigraphy of valley fills at Aeolis Dorsa, Mars: evidence for base-level fluctuations controlled by a downstream water body. GSA Bull 130(3–4):484–498

Catuneanu O, Galloway WE, Kendall CGSC, Miall AD, Posamentier HW, Strasser A, Tucker ME (2011) Sequence stratigraphy: methodology and nomenclature. Newsl Stratigr 44(3):173–245

Flood RC, Manley PL, Kowsmann RO, Appi J, Pirmez C (1991) Seismic facies and late quaternary growth of the Amazon submarine fan. In: Weimer P, Links MH (eds) Seismic facies and sedimentary processes of modern and ancient submarine fans and turbidite systems. Springer, New York, pp 247–272

Grimaud JL, Paola C, Voller V (2016) Experimental migration of knickpoints: influence of style of base-level fall and bed lithology. Earth Surf Dyn 4(1):11–23

Heller PL, Paola C, Hwang I, John B, Steel R (2001) Geomorphology and sequence stratigraphy due to slow and rapid base-level changes in an experimental subsiding basin (XES 96-1). Am Assoc Pet Geol Bull 85:817–838

Holbrook J, Scott RW, Oboh-Ikuenobe FE (2006) Base-level buffers and buttresses: a model for upstream versus downstream control on fluvial geometry and architecture within sequences. J Sediment Res 76(1):162–174

Kolla V, Perlmutter MA (1993) Timing of turbidite sedimentation of the Mississippi Fan. Am Assoc Pet Geol Bull 77:1129–2114

Leopold LB (1992) Base-level rise: gradient of deposition. Israel J Earth Sci 41:57–64

Leopold LB, Bull WB (1979) Base-level, aggradation and grade. Proc Am Philos Soc 123:168–202

Leopold LB, Wolman MG, Miller JP (1964) Fluvial processes in geomorphology. WH Freeman and Company, San Francisco, p 507

Muto T, Steel RJ (2002) In defense of shelf-edge delta development during falling and low stand of relative sea level. J Geol 110(4):421–436

Parker G, Akamatsu Y, Muto T, Dietrich W (2004) Modeling the effect of rising sea level on river deltas and long profiles of rivers

Schumm SA (1993) River response to base-level change: implications for sequence stratigraphy. J Geol 101(2):279–294

Schumm SA (2005) River variability and complexity, chap 16. Cambridge University Press, Cambridge, pp 149–155

Vail PR, Mitchum RM Jr, Thompson S III (1977) Seismic stratigraphy and global changes of sea level: part 3. Relative changes of sea level from Coastal Onlap. Mem Am Assoc Pet Geol 26:63–81

Controlling Factors

6

Abstract

The following key factors control the impact of the base-level: 1. The *contributing drainage area* which is the proxy for sediment and water discharge that controls and regulates the erosion efficiency. Knickpoints propagate upstream at a rate characterized by the power law function of the upstream drainage area. 2. The *rock strength* affects the erodibility and includes the dip and its orientation, jointing and layering or massiveness. 3. The *sediments* which provide an incision tool and might later form a sediment armor that protects the streambed until flushed out. The ratio of sediment supply to transport capacity represents competing effects. 4. The *coupling between hillslopes and channels.* Onset of landsliding provides a large source of new and coarse sediments. 5. The *slope gradient* exposed by the base-level fall is critical in determining the response to the base-level drop. Entrenchment is likely to occur only if the exposed gradient exceeds the channel bed gradient to which the river would normally adjust at grade. 6. The *magnitude, rate and duration* of the base-level change. In case base-level lowering is small, the channel may adjust only by increased sinuosity. The longer the duration of a base-level fall, the more adjustments of the fluvial system can be expected. Because of different background conditions as climate, lithology and tectonics, fluvial systems may respond differently to the same base-level change.

Keywords

Substrate • Drainage area • Erosion efficiency • Sediment armor • Complex response • Allogenic/Autogenic factors • Unconformity type 1 and 2 • Hillslopes-channel coupling

We differentiate between *allogenic* factors, which control systems from outside (external factors) and *autogenic* dynamics that control systems from within (internally generated processes) that produce *self-organization*, i.e., a regime whereby the overall order is based on local interactions and feedbacks between parts *within*

D. Bowman, *Base-level Impact*,
https://doi.org/10.1007/978-3-031-24994-5_6

the system that are able to self-repair perturbations. The drainage network, of which the base-level is a major component, is as all geomorphic structures are, an *open system*, i.e., exposed not only to internal interactions but to factors from outside as well. The key factors, controlling the impact of the base-level, are the following:

6.1 The Contributing Drainage Area

The upstream contributing drainage area is the first-order control of the sediment and water supply to the channel, making it a proxy for both and a regulator of the erosional efficiency (Leopold and Miller 1956; Leopold et al. 1964). The catchment with its annual precipitation, rainfall intensity and duration, and the flood magnitude and frequency control the knickpoint retreat rate and regulates the lateral migration and vertical incision. However, although water discharge is a primary factor that determines the rate of profile adjustment, numerous additional competing controls make it difficult to definitively conclude the exact role of the drainage basin area alone (Crosby and Whipple 2006).

When stream discharge is small the response to a lowering base-level will be slower and equilibrium takes longer to attain. Fast profile adjustment is observed in streams with a larger catchment area. Changes in the drainage area may also occur following capture, when drainage divisions are breached following base-level fall (Kooi and Beaumont 1996; Fig. 2.3). In case of divide migration triggered by a lower and more effective base-level, stored sediments in the captured tributary could be flushed while downstream reaches of the now dominating drainage system are still graded for transporting a former, low-sediment discharge.

There is a highly significant relationship between the distance of knickpoint recession and the catchment area (Bishop et al. 2005). The knickpoint propagates upstream at a rate that is a power law function of the upstream drainage area (Wobus et al. 2006; Loget and Van Den Driessche 2009), i.e., at an ever-decreasing rate that is proportional to the gradual decrease in the contributing drainage area and the consequently decreasing stream power. Knickpoints at the same topographic altitudes, in different tributaries, even of variable lithologies, may suggest a similar pulse of incision activated by a similar size of drainage basin (Fig. 6.1).

Based on the conception that the drainage area is a proxy for both sediment and water discharge (Leopold and Miller 1956; Leopold et al. 1964), the upstream knickpoint retreat, proportional to the declining watershed area, causes the knickpoint recession rate to decline through time (Schumm et al. 1987). Modeled knickpoints that migrated quickly when activated by large drainage areas significantly slowed down when activated by smaller drainage areas. The longer the time duration the lower became the migration rate. Experiments by Brush and Wolman (1960) and by Leopold et al. (1964) showed an exponential migration curve of

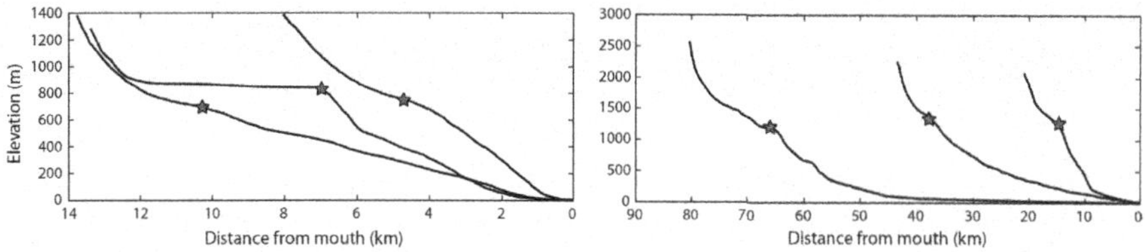

Fig. 6.1 Examples of knickpoints (stars) in adjacent streams at similar elevations, eastern Papua New Guinea. All knickpoints resulting from the same perturbation in uplift rate under similar erodibility conditions should be found on similar topographic elevation. Knickpoints that occur in relatively narrow elevation bands, where rock uplift (base-level fall) rates and stream erosion processes are uniform, without similar subhorizontal stratigraphy, may indicate transient migration at a similar rate (From Miller et al. 2012)

knickpoints with the length of the equilibrated reach (Fig. 6.2). The travel distance and the celerity of knickpoints have been suggested as power functions of the drainage area with an exponent between 0.45 and 0.55 (Beckers et al. 2015).

Larger catchments are expected to generate higher sediment volumes and thus have a more rapid response. High-order drainage systems have a better chance to adjust to base-level drop, whereas lower-order streams will keep steeper (Merritts

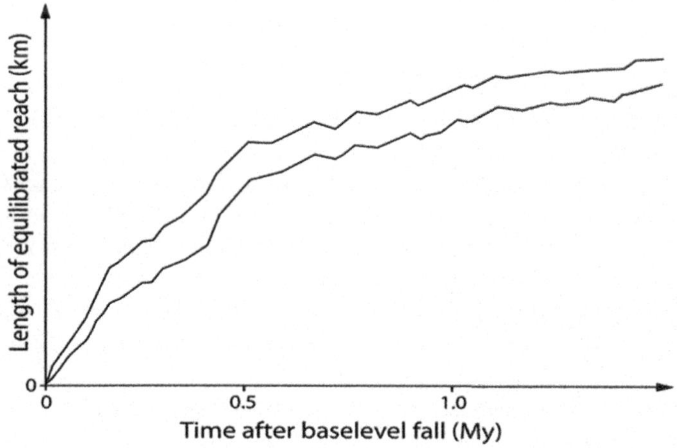

Fig. 6.2 Model evolution of the length of an equilibrated reach through time. The length of channels affected by base-level fall increases with time, whereas the migration rate decreases. An envelope of high and low rates of propagation is indicated (Modified from Attal et al. 2008) Apart from the base-level lowering rate as a key variable that sets the retreat, knickpoint migration is primarily drainage-area-dependent (Crosby and Whipple 2006) modulated by bedrock lithology (Valla et al. 2010). Larger catchments are expected to generate higher water and sediment volumes and a far-distant retreat of knickpoints up the fluvial network. Gradual absence of erosion tools, i.e., bedload starvation, explains the decrease of the knickpoint retreat rate. Early in the knickpoint retreat process, the rates of knickpoint retreat are the fastest but will become slower as the contributing drainage area decreases

and Vincent 1989). Knickpoints are more common on first-order streams, and their frequency decreases with increase in stream order and power.

Transmission of the base-level fall signal by a propagating knickpoint requires a minimum value of discharge. A smaller than *critical drainage area* does not support the needed discharge of water and sediment, leaving the channel in a *detachment–limited* status. Small drainage areas are unable to respond to base-level fall. Their limited stream power renders them incapable of substantially incising.

6.2 The Substrate

When it comes to knickpoint dynamics, the effect of the substrate, including its structure and sediment cover, may not be the first-order control (Sklar and Dietrich 1998). According to this concept, the substrate may introduce local differences among channels but does not seem to affect knickpoint retreat overall (Castillo et al. 2013). Increasing rock strength is favorable for creation of knickpoints whose upstream propagation is slower and may even cause knickpoints to become anchored by hard lithology. Of importance is as well the dip and its orientation with respect to the hillslope orientation. Dipping toward the hillslope or out of it controls the mechanical stability of slopes. Moreover, dominant subvertical jointing may exert a profound control on the evolution of gorges' headwalls and equally on the adjacent sidewalls and enhance their retreat rate. When the rate of upstream knickpoint migration is strongly controlled by jointing and fracturing, the fluvial processes although important may play only a secondary role (Weissel and Seidl 1997, 1998).

The importance of erodibility can be demonstrated by variations in the bedrock incision rate along streams which may dictate the location of the knickpoints. Resistant formations may cause the incision rate to drop below the base-level lowering rate. The resistance of a caprock relative to the strength of the underlying rock and the critical face heights for gravitational failure also control the retreat rate of knickpoints. Massive caprocks tend to have deeper overhangs relative to knickpoints whose caprock formation has closely spaced horizontal bedding. A layered caprocks may not be able to support an abrupt overhang and thus tend to become stepped knickpoints (Haviv et al. 2010). The effect of erodibility may be higher in alluvial rivers than in bedrock rivers where low mean rates of knickpoint migration are apparent.

6.3 The Sediment Flux

The overall result of a base-level drop is an upward channel expansion and a simultaneous downstream movement of sediments. Knickpoint migration headwards and sediment discharge at the outlet are strongly related (Parker 1977). Sediments are supplied from the receding headcuts, from the banks, the hillslopes and from entrenchment of the channel and are a certain proxy of the supplying drainage area. (Sklar and Dietrich 2004; Whipple and Tucker 1999; Turowski et al. 2007). The sediments moving along channels may erode the bed or cover and protect it. At the initial stage, the sediments provide efficient incision tools that drive the knickpoint retreat rates but later may form a sediment cover along the channel, i.e., an armor of lag deposits which increases the streambed resistance and protects it by increasing the hydraulic roughness which stabilizes the bed and prevents further degradation until flushed out. The bottom of a bedrock gorge is periodically covered by such lag sediments which would be subsequently evacuated. After the initial incision stage, sediment yield decreases as an intrinsic evolution of the system. Part of the diminishing sediment load reflects increasing sediment storage. Secondary peaks during the declining sediment yield reflect post-storage flushing events.

The particle size of the sediment has a strong influence on the channel slope following base-level fall. In laboratory studies, changes of the slope in finer material were much greater than in coarser (Brush and Wolman 1960). Retreat velocity of knickpoints will decrease with boulder size supplied from the walls because of insufficient water discharge to mobilize and evacuate such large sediment. Gradual depletion of paraglacial sediment supply over the Holocene lead to a deficiency in tools which drove down erosion rates (Sklar and Dietrich 2001).

The ratio of sediment supply to transport capacity represents competing effects, i.e., eroding the bed or covering and protecting it. When sediment flux exceeds transport capacity, sediment will cover and armor the bed against further incision and force the channel toward a transport-limited behavior. The stream power incision rule assumes that transport capacity well exceeds the imposed sediment load and thus the rate of channel incision is limited simply by the channel's capacity to detach bedrock from the channel bed (Howard and Kerby 1983; Whipple and Tucker 1999).

We expect correlation between headwall migration distance and stream discharge (Hayakawa and Matsukura 2003; Crosby and Whipple 2006). Enhanced sediment supply would too drive faster knickpoint retreat. The greater the distance from an outlet headwards, the smaller the sub-basin nourishing area becomes and smaller will be the runoff and the sediment supply, all of which contribute to decreasing rates of upstream knickpoint propagation (Parker 1977). Maximum incision rates occur at moderate sediment supply rates relative to the sediment transport capacity, because of a tradeoff between the availability of abrasive tools and the partial alluviation of the streambed. Lower water discharges and high sediment concentration result in a greater lag time of response, confirmed across a range of sediment sizes (Bonneau and Snow 1992). Warmer and wetter climates that increase vegetation cover and provide greater soil cohesion, reduce sediment supply. As knickpoints prograde upstream and pass a site, sediment yield increases notably to maximum and thereafter decays exponentially.

In headwater areas, sediment loads which cannot be transported immediately, cause local aggradation and form storage of alluvial fill. Later, after being periodically stored, the sediment will be flushed out (Parker 1977). Such evolution presents a *complex response,* i.e., an analog of Schumm's "*complex response*" concept (Schumm 1977; Schumm and Rea 1995) in which geomorphic systems respond in complex ways to external changes. Based on experiments, Schumm et al. (1984) indicated that the response to rejuvenation by base-level fall is complex in short periods, demonstrated by episodes of incision followed by aggradation, but in the long run clear response trends dominate.

6.4 Slope Sourcing

In exploring the response to base-level fall, we should include sediments released down hillslopes (Whittaker et al. 2010). The incision due to base-level fall develops close coupling between hillslopes and channels (Fig. 6.3). In regions of high relief, slopes are typically maintained at or near the threshold failure angle (~30°). Beyond threshold, larger volumes of material are delivered from hillslopes, nourishing the alluvial cover that armors the channel bed. Such hillslope activity typically lags behind channel dynamics. Influx of large, immobile blocks from channel-adjacent hillslopes inhibit downcutting by directly shielding the bed from erosion. In a block flux experiment, retreat rates became >50% slower on average (Shobe et al. 2016). Reaches of mountain rivers that host clusters of large (>1 m) blocks slow down the propagation of knickpoints up the drainage system and thereby form a negative feedback.

Sediments derived from the valley slopes and form talus piles constrain the channel width. The shear stress and the sediment transport capacity per unit width of narrowing channels increase and the channels tend strongly to incise into its bed. Narrowing of channels by lateral inputs of talus material forms a *positive feedback* eroding loop. Banks that constrain the channel and reduce its width enhance a greater vertical incision (Malatesta et al. 2017).

The grain size signal that is released through landslides and scree cones directly into incising channels is substantially coarser than the ones supplied by the channel from upstream. Larger and more frequent landslides are expected following

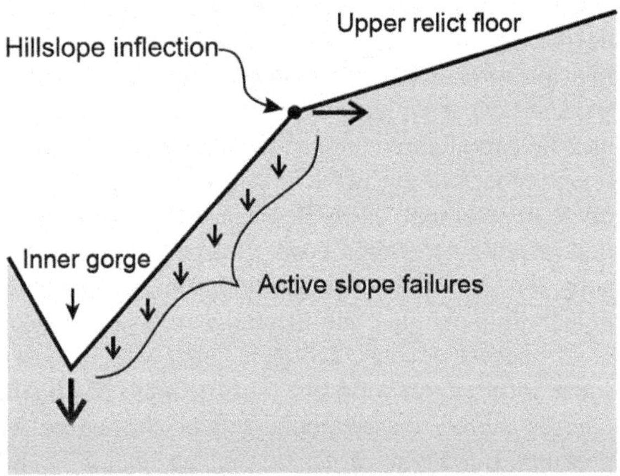

Fig. 6.3 Schematic cross section of a valley highlighting the steep slopes of an *inner gorge* entrenched in the valley. Landslides and scree cones that were maintained near the threshold of failure on the valley slope, prior to the base-level fall, become activated by incision of the *inner gorge*, triggered by a falling base-level. The incised inner gorge will gradually dominate the entire cross section (Modified after Reinhardt et al. 2007)

larger base-level fall events that drive deeper incision. Onset of landsliding provides a large source of new coarse sediments *in addition* to the material being supplied from the upstream catchment. Coarse sediments must not be related to the upper drainage area, thus preventing studies to attempt and derive catchment wide erosion rates (Whittaker et al. 2010). In interpreting base-level impacts in mountain rivers, autogenic channel-hillslope interactions which supply hillslope-derived blocks in response to the river incision should be included. The coarse landslides material will affect the depositional stratigraphy and will be detectable in the proximal hanging wall basin stratigraphy by coarsening-up sequences.

6.5 Slope Gradients

Field evidence and laboratory data show that the *initial slope gradient*, exposed following base-level fall, is critical in determining the response to the base-level drop (Talling 1998). Channel incision rate is linearly dependent on the initial slope gradient and increases accordingly. Continental margins serve as a good example to the slope effect. The shelf or the continental slope, when exposed, determine the trends which accompany eustatic fluctuations. *Type 1 unconformity* of Posamentier and Vail (1988) represents a fall of base-level upon a steep slope, beyond the shelf break, triggering deep incision. *Type 2 unconformity* marks a base-level lowering across a subtle shelf slope, exposing wide areas without activating intensive entrenchment. Hence, in response to the same base-level fall, incision *will not* be alike. Rivers *will not* incise into their substrate during base-level fall if the exposed area is of low gradient and therefore insufficient to trigger incision. Entrenchment is only likely to happen if the exposed gradient exceeds the channel bed gradient to which the river would normally adjust at grade in a given water and sediment supply regime (Miall 1991; Posamentier et al. 1992; Schumm 1993). Even modest base-level falls can induce incision when the exposed area is steep enough. When, following base-level fall, the gradient of the exposed area is low, the response may even keep an aggradational trend if the regime is of high sediment delivery rate.

The steeper the slope, the faster the wave of re-equilibration propagates upstream. As water tends to flow faster in steeper reaches and therefore occupy smaller channel cross sections, an increase in channel slope leads to reduction of the channel width. Slopes are, in addition, proportional to a function of the particle size of the bedload and tend to decrease exponentially in downstream direction. The efficiency of river incision by bedload abrasion decreases with increasing gradients *beyond* a certain threshold, delaying headward incision and thus preserving, for example, hanging valleys (Sklar and Dietrich 2004).

6.6 Magnitude and Rate

The effect of a base-level fall on the drainage basin depends on the *rate* and *magnitude* of the change (Yoxall, 1969; Schumm, 1993; Heller et al. 2001). Dating of marine or lacustrine terraces that indicate the altitude of former base-levels (see front cover) provides the rate of the base-level fall. When lowering of base-level is fast, streams incise vertically with little lateral migration. When the rate is slow the impact will be smaller (Schumm et al.1984) with a considerable lateral migration induced (Yoxall 1969; Wood et al. 1993; Schumm 1987). The deepest, steepest, and narrowest incised valleys with the most significant upward evolution are formed when base-level fall is most rapid (Begin et al. 1981; Schumm 1993; Blair and Mcpherson 1994). Schumm (1977) showed that after increasing and reaching peak, the rate of degradation at any point along the channel slowly decreases as the channel becomes gradually adjusted to the new base-level. Main erosion is felt in the initial stages of the base-level lowering and mainly near the outlet.

6.7 Response Variability and Complexity

Rivers may respond to changes in base-level not only by incision or aggradation. Small-amplitude base-level changes can be absorbed by internal adjustments in the fluvial system which counteract the effect of external changes. Channels may adjust by widening, increase in roughness and decrease in depth, thereby lower the velocity and the stream power (Leopold and Bull 1979). Channels may also adjust by changing the channel pattern (Schumm 1977, 1993, 2005) which may, however, absorb only part of the base-level impact (Leopold and Bull 1979). Channels may, for example, adjust by increasing sinuosity close to the base-level (Schumm et al. 1987), thereby maintain the channel gradient in spite of steepening. However, an increased sinuosity can only partly compensate for steepening of a valley floor. Dente et al. (2018) reported an increase of sinuosity in downstream channel sections of the Jordan River in response to the fall of the Dead Sea level. Meanders as a response to base-level fall have been demonstrated as well by tributaries of the Yangtze River in China. The tributaries may represent the response to the Quaternary fall of the Yangtze River as their local base-level (Li et al. 2001).

Extending the time of base-level fall lengthens the duration of incision and will diminish the incision rate. By distributing the same magnitude of base-level fall over a longer period of time, tributary junctions have an improved probability of keeping pace with the mainstream incision, resulting in a lower likelihood of creating hanging valleys (Crosby 2006).

Channel networks may be regulated simultaneously by the different controlling factors (Leopold and Bull 1979; Currie 1997; Muto and Steel 2004; Hayakawa and Matsukura 2003). Hence, although base-level lowering may be identical, vertical incision rate may vary widely reflecting the other various factors controlling the entrenchment process (Hancock and Anderson 2002). In addition, *Upper controlling factors* active from the divides downstream may be out of phase with factors controlling the system from downstream up. For example, deglaciation may trigger fluvial entrenchment in upstream reaches while sea level rise controls the downstream areas (Catuneanu et al. 2009), demonstrated in the Mattole and Bear Rivers in northern California, where down cutting continued during the recent sea level rise (Merritts et al. 1994).

Due to different relief and climate conditions, different lithology, tectonic movements, sediment supply and different stream discharge, the fluvial systems may respond differently to the same base-level change (Martinsen et al. 1999). Delta aggradation may thus persist in spite of a sea level fall when its high sediment supply is sufficient to prevent incision. Hence, global eustatic changes do not provide a great unifying generalization and do not permit worldwide correlation of erosion surfaces (Schumm 1993). In addition, there are internal fluvial processes that have the same degradational effect as base-level lowering. For example, lateral migration of a trunk stream at a tributary junction toward the branch, shortens the tributary length and has a local base-level lowering effect. Controlling factors and responses to base-level changes are variable and complex. We should therefore conclude that assigning allogenic controls to changes in morphological trends should be carried out thoroughly.

References

Attal M, Tucker GE, Whittaker AC, Cowie PA, Roberts GP (2008) Modeling fluvial incision and transient landscape evolution: influence of dynamic channel adjustment. J Geophys Res: Earth Surf 113(F3)

Beckers A, Bovy B, Hallot E, Demoulin A (2015) Controls on knickpoint migration in a drainage network of the moderately uplifted Ardennes Plateau, Western Europe. Earth Surf Proc Land 40(3):357–374

Begin ZB, Meyer DF, Schumm SA (1981) Development of longitudinal profiles of alluvial channels in response to base-level lowering. Earth Surf Proc Land 6(1):49–68

Bishop P, Hoey TB, Jansen JD, Artza IL (2005) Knickpoint recession rate and catchment area: the case of uplifted rivers in Eastern Scotland. Earth Surf Proc Land 30(6):767–778

Blair TC, McPherson JG (1994) Alluvial fans and their natural distinction from rivers based on morphology, hydraulic processes, sedimentary processes, and facies assemblages. J Sediment Res 64(3a):450–489

Bonneau PR, Snow RS (1992) Character of hearwaters adjustment to base-level drop, investigated by digital modeling. Geomorphology 5(3–5):475–487

Brush LM, Gordon Wolman M (1960) Knickpoint behavior in noncohesive material: a laboratory study. Geol Soc Am Bull 71(1):59–74

Castillo M, Bishop P, Jansen JD (2013) Knickpoint retreat and transient bedrock channel morphology triggered by base-level fall in small bedrock river catchments: the case of the Isle of Jura, Scotland. Geomorphology 180:1–9

Catuneanu O, Abreu V, Bhattacharya JP, Blum MD, Dalrymple RW, Eriksson PG, Giles KA (2009) Towards the standardization of sequence stratigraphy. Earth Sci Rev 92(1–2):1–33

Crosby BT (2006) The transient response of bedrock river networks to sudden base-level fall. Doctoral dissertation, Massachusetts Institute of Technology

Crosby BT, Whipple KX (2006) Knickpoint initiation and distribution within fluvial networks: 236 waterfalls in the Waipaoa River, North Island, New Zealand. Geomorphology 82(1–2):16–38

Currie BS (1997) Sequence stratigraphy of nonmarine Jurassic-Cretaceous rocks, central Cordilleran foreland-basin system. Geol Soc Am Bull 109(9):1206–1222

Dente E, Lensky NG, Morin E, Dunne T, Enzel Y (2018) Sinuosity evolution along an incising channel: New insights from the Jordan River response to the Dead Sea level fall. Earth Surf Proc Land 44(3):781–795

Hancock GS, Anderson RS (2002) Numerical modeling of fluvial strath-terrace formation in response to oscillating climate. Geol Soc Am Bull 114(9):1131–1142
Haviv I, Enzel Y, Whipple KX, Zilberman E, Matmon A, Stone J, Fifield KL (2010) Evolution of vertical knickpoints (waterfalls) with resistant caprock: insights from numerical modeling. J Geophys Res: Earth Surf 115(F3)
Hayakawa Y, Matsukura Y (2003) Recession rates of waterfalls in Boso Peninsula, Japan, and a predictive equation. Earth Surf Proc Land 28(6):675–768
Heller PL, Paola C, Hwang IG, John B, Steel R (2001) Geomorphology and sequence stratigraphy due to slow and rapid base-level changes in an experimental subsiding basin. Bull Am Assoc Pet Geol 85:817–883
Howard AD, Kerby G (1983) Channel changes in badlands. Geol Soc Am Bull 94:739–752
Kooi H, Beaumont C (1996) Large-scale geomorphology: classical concepts reconciled and integrated with contemporary ideas via a surface processes model. J Geophys Res: Solid Earth 101(B2):3361–3386
Leopold LB, Bull WB (1979) Base-level, aggradation and grade. Proc Am Philos Soc 123:168–202
Leopold LB, Miller JP (1956) Ephemeral streams-hydraulic factors and their relation to the drainage net. USGS Professional Paper 282-A: 1–3
Leopold LB, Wolman MG, Miller JP (1964) Fluvial processes in geomorphology. W.H. Freeman and Co, San Francisco, 522p
Li J, Xie S, Kuang M (2001) Geomorphic evolution of the Yangtze Gorges and the time of their formation. Geomorphology 41(2–3):125–135
Loget N, Van Den Driessche J (2009) Wave train model for knickpoint migration. Geomorphology 106(3–4):376–382
Malatesta LC, Prancevic JP, Avouac JP (2017) Autogenic entrenchment patterns and terraces due to coupling with lateral erosion in incising alluvial channels. J Geophys Res Earth Surf 122(1):335–355
Martinsen OJ, Ryseth AL, Helland-Hansen WI, Flesche H, Torkildsen G, Idil S (1999) Stratigraphic base-level and fluvial architecture: Ericson Sandstone (Campanian), Rock Springs Uplift, SW Wyoming, USA. Sedimentology 46(2):235–263
Merritts D, Vincent KR (1989) Geomorphic response of coastal streams to low, intermediate, and high rates of uplift, medocino tripple junction region, Northern California. Geol Soc Am Bull 101(11):1373–1388
Merritts DJ, Vincent KR, Wohl EE (1994) Long river profiles, tectonism, and eustasy: a guide to interpreting fluvial terraces. J Geophys Res: Solid Earth 99(B7):14031–14050
Miall AD (1991) Stratigraphic sequences and their chronostratigraphic correlation. J Sediment Petrol 61:497–505
Miller SR, Baldwin SL, Fitzgerald PG (2012) Transient fluvial incision and active surface uplift in the Woodlark Rift of eastern Papua New Guinea. Lithosphere 4(2):131–149
Muto T, Steel RJ (2004) Autogenic response of fluvial deltas to steady sea-level fall: implications from flume-tank experiments. Geology 32(5):401–404
Parker RS (1977) Experimental study of basin evolution and its hydrologic implication: unpublished Ph.D. dissertation, Colorado State University, Port Collins, Colorado
Posamentier HW, Allen GP, James DP, Tesson M (1992) Forced regressions in a sequence stratigraphic framework: concepts, examples, and exploration significance. Am Assoc Petroleum Geol Bull 76:1687–1709
Posamentier HW, Vail PR (1988) Eustatic controls on clastic deposition. II. Sequence and systems tract models. In: Wilgus CK, Hastings BS, Kendall CGSTC, Posamentier HW, Ross CA, Van Wagoner JC (eds) Sea level changes—an integrated approach. SEPM Special Publication, vol 42, pp 125–154
Reinhardt LJ, Bishop P, Hoey TB, Dempste TJ, Sanderson DCW (2007) Quantification of the transient response to base-level fall in a small mountain catchment: Sierra Nevada, Southern Spain. J Geophys Res: Earth Surf 112(F3)
Schumm SA (1977) An experimental study of geomorphic thresholds. Colorado State University, Fort Collins

Schumm SA (1993) River response to base-level change: implications for sequence stratigraphy. J Geol 101:279–294

Schumm SA (2005) River variability and complexity. Cambridge University Press, Cambridge, Chap 16, pp 149–155

Schumm SA, Harvey MD, Watson CC (1984) Incised channels: morphology, dynamics, and control. Water Resources Publications

Schumm SA, Mosley MP, Weaver W (1987) Experimental fluvial geomorphology. Wiley, New York, p 413

Schumm SA, Rea DK (1995) Sediment yield from disturbed earth systems. Geology 23(5):391–394

Shobe CM, Tucker GE, Anderson RS (2016) Hillslope-derived blocks retard river incision. Geophys Res Lett 43(10):5070–5078

Sklar L, Dietrich WE (1998) River longitudinal profiles and bedrock incision models: stream power and the influence of sediment supply. Geophys Monograph-Am Geophys Union 107:237–260

Sklar LS, Dietrich WE (2001) Sediment and rock strength controls on river incision into bedrock. Geology 29(12):1087–1090

Sklar LS, Dietrich WE (2004) A mechanistic model for river incision into bedrock by saltating bed load. Water Resources Res 40(6)

Talling PJ (1998) How and where do incised valleys form if sea level remains above the shelf edge? Geology 26(1):87–90

Turowski JM, Lague D, Hovius N (2007) Cover effect in bedrock abrasion: a new derivation and its implications for the modeling of bedrock channel morphology. J Geophys Res: Earth Surf 112(F4)

Valla PG, van der Beek PA, Lague D (2010) Fluvial incision into bedrock: insights from morphometric analysis and numerical modeling of gorges incising glacial hanging valleys (Western Alps, France). J Geophys Res 115

Weissel JK, Seidl MA (1997) Influence of rock strength properties on escarpment retreat across passive continental margins. Geology 25(7):631–634

Weissel JK, Seidl MA (1998) Inland propagation of erosional escarpments and river profile evolution across the Southeast Australian passive continental margin. Geophys Monograph-Am Geophys Union 107:189–206

Whipple KX, Tucker GE (1999) Dynamics of the stream-power river incision model: implications for height limits of mountain ranges, landscape response timescales, and research needs. J Geophys Res: Solid Earth 104(B8):17661–17674

Whittaker AC, Attal M, Allen PA (2010) Characterising the origin, nature and fate of sediment exported from catchments perturbed by active tectonics. Basin Res 22(6):809–828

Wobus C, Whipple KX, Kirby E, Snyder N, Johnson J, Spyropolou K, Crosby B, Sheehan D (2006) Tectonics from topography: procedures, promise, and pitfalls. Special papers-Geological Society of America 398, Penrose Conference Series, pp 55–74

Wood LJ, Ethridge FG, Schumm SA (1993) An experimental study of the influence of subaqueous shelf angles on coastal plain and shelf deposits: Chapter 15: recent developments in siliciclastic sequence stratigraphy

Yoxall WH (1969) The relationship between falling base-level and lateral erosion in experimental streams. Geol Soc Am Bull 80:1379–1384

7 Knickpoint Retreat

Abstract

The base-level fall signal is transmitted through the drainage system as a transient (migrating upstream) erosional wave in form of a knickpoint or knickzone. Its movement includes: (1) migrating backwards and (2) moving vertically. There is no one-to-one correlation between knickpoints along river profiles and base-level events. Knickpoint retreat scales with stream discharge and gradient. The travel distance and the celerity of knickpoints have been suggested as power functions of the drainage area. All experiments show a delay between the base-level fall and the appearance of erosional features upstream. The greater the drainage area and the higher the rate of the base-level fall, the shorter is the expected delay. Additional competing controls that determine the rate of knickpoint retreat include the stream gradient, weathering processes, abrasion and undermining.

Keywords

Transient wave • Waterfall • Plunge pool • Complex response • Degradation index • Diachroneity • Lag

Base-level lowering shifts the locus of incision upslope, forming a migrating knickpoint, i.e., a transient responses to lowering. This bottom-up process is the communication link between the base-level and the entire catchment including hillslopes. Knickpoint migration communicates as a kinematic wave (Wobus 2005) through the entire drainage net, including the trunk channel and its tributaries (Fig. 7.1). Hence, knickpoint retreat is the mechanism through which incision propagates upstream and extends as a primary mode of landscape evolution (Bishop et al. 2005). When the knickpoint migrates upstream and controls larger parts of the drainage system it possibly crosses variable climatic, lithologic and tectonic zones, each one contributing its specific impact. This is spatially very different from a climatic change that affects the entire drainage system and may have a more sudden impact (Bull 1991).

D. Bowman, *Base-level Impact*,
https://doi.org/10.1007/978-3-031-24994-5_7

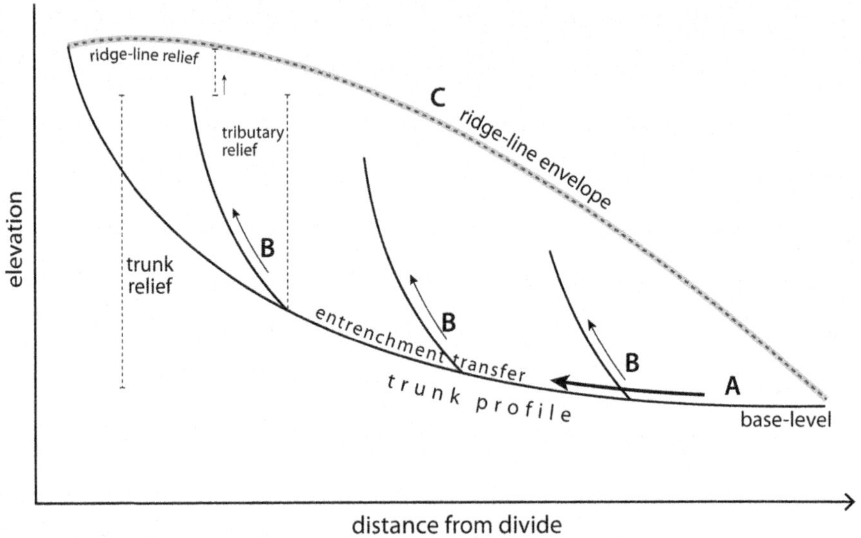

Fig. 7.1 Incision and relief production along a fluvial drainage network in response to base-level fall (leaving out tectonic uplift and glaciation). Entrenchment is transferred upstream through the alluvial trunk channel (A). The upstream transient incisional wave carries on along the steepening tributaries (B) and finally dies out toward the upper reaches and the ridge line (C) (schematic, modified after Whipple et al. 1999)

A single base-level fall may form a single knickpoint. Multiple knickpoints may indicate several base-level fall events. However, self-organization may form a single knickpoint out of multiple small base-level steps (Grimaud et al. 2016). There is no one-to one correlation between knickpoints along river profiles and base-level events: One base-level drop can generate multiple knickpoints and one knickpoint can result from multiple events. An upstream migrating single knickpoint may transform into a series of small knickpoints that can develop and die out during upstream migration (Begin 1978; Schumm et al. 1984). A series of knickpoints caused by various base-level events may merge at one resistant section along the stream bed. A large number of knickpoints that migrate upstream, related to a single lowering base-level, suggest that the base-level may not be their only control. When cluster at a similar altitude (Fig. 6.1) in different networks, knickpoints may indicate a common base-level origin (Robl et al. 2017) and a similar vertical celerity of migration.

7.1 Waterfall Knickpoint Retreat

Waterfalls often serve as examples for knickpoints. Their withdrawal involves migrating backwards and vertically up the channel profile. Knickpoint propagation is thought to be controlled by the river stream power (Brocard et al. 2016;

Crosby and Whipple 2006) or the basal shear stress. Both are often used as proxies for estimating fluvial erosion. In models, knickpoint retreat scales with stream discharge and stream gradient (Bishop et al. 2005; Crosby and Whipple 2006). Several studies suggest that knickpoint retreat is more sensitive to drainage area than to river gradient (Bishop et al. 2005). Some weakness of the dependency of knickpoint celerity on the drainage area can be shown by the sensitivity of the knickpoint retreat to joint orientation as well as to weathering processes such as exfoliation, freeze–thaw and wet-dry-related weathering cycles and to groundwater seepage and to toppling of large blocks.

The actual shear stress (τ_o), i.e., that is produced by the fluid flow, compared to the critical shear stress (τ_c) that is needed for erosion to take place, attains a maximum value at the sharpest gradient change along the knickpoint (Gardner 1983; Schumm et al. 1987). Leopold et al. (1964) predicted that knickpoints would migrate upstream, evolve or diffuse, if stream flow is competent to transport material downstream and if the ratio between the height of the knickpoint face (H_k) and flow depth (h) is greater than one ($H_k/h > 1$; Fig. 7.2). Peak stream discharge must exceed a threshold sufficient to transport gravel and debris away, keeping the knickpoint face free, thus allowing channel incision. The headcut height decreases when the deposition of debris outpaces the fluvial transport of material away from the head. Abrasion and plucking on the exposed bedrock channel floor above the knickpoint drive vertical incision and gradually reduce the height of the knickpoint face. Cohesive sediments can help a propagating knickpoint to preserve its shape, preventing diffusion with distance and allowing it to progress farther upstream.

Significant shear stress and abrasion are effectively accomplished by sediment impacts and the direct falling water jet at contact with the waterfall face (Fig. 7.3). The shear stresses imposed by falling water activate abrasion and plucking from the waterfall face and might be particularly important for bedrock erosion in well-fractured rocks (Lamb et al. 2007). When the incision rate headwards becomes significantly less than the rate of base-level fall, the knickpoint will steepen. An

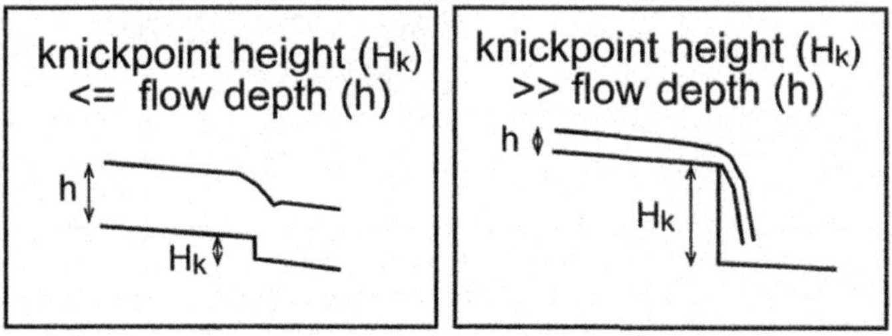

Fig. 7.2 Knickpoints of different ratios between height of the knickpoint face and flow depth. When knickpoint height < flow depth the knickpoint becomes submerged. When knickpoint height ≫ flow depth a waterfall will form and the knickpoint face will rereat ontrolled by gravitational failure, direct abrasion, sediment impacts, water jet impact, freeze and thaw cycles, weathering and by the plunge pool effect (from Haviv et al. 2010)

increase in waterfall height can be attained as long as the vertical incision rate below the waterfall is greater than upstream of the waterfall (Haviv et al. 2010).

A very steep knickpoint face enables very little contact between the falling water jet and the knickpoint face. The shear stress exerted from flowing water reaches a maximum at inclinations of 45ø and diminishes to zero with vertical free-fall conditions. The decreasing erosion rates are due to increasing particle hop lengths (DiBiase et al. 2015). As the efficiency of kinetic energy transfer to a bedrock surface decreases at high channel gradients (Sklar and Dietrich 2004),

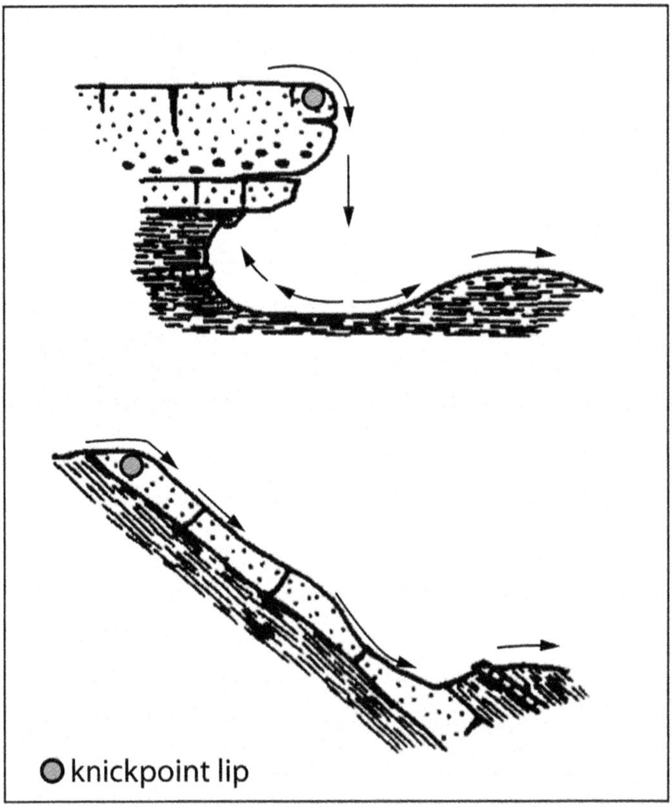

Fig. 7.3 **Upper sketch** demonstration of plunge pool erosion, when water falling into the pool from a vertical scarp (headcut) forms a jet divided in the pool into two flows parallel to the bed, (1) scouring in the downstream direction, where the maximum shear stress occurs and (2) producing a flow reversal in the upstream direction in form of a backroller that entrains material toward the face of the headcut, eroding its base if free and not protected by debris (Bressan et al. 2014; Flores-Cervantes et al. 2006). Competition prevails between reducing the headcut height by the overland flow and scouring in the plunge pool. The headcut is maintained when the retreat rate is faster than overland flow erosion. **Lower sketch** demonstrates a knickpoint on a steep dipping structure (after Alexandrowicz 1994)

channels with very steep gradients erode at a lower rate. The retreat of waterfalls requires other processes, such as plunge pool undermining, where weathering and fracturing play a significant role (Howard et al.1994). Thus, at steep gradients knickpoint migration rate is no longer a simple function of the upstream drainage area only. However, there are observations of rapidly retreating waterfalls at odds with the bedload saltation-abrasion incision models, which predict decreasing erosion rates due to increasing particle hop lengths (DiBiase et al. 2015).

Undermining is an important process in retreat of many waterfall knickpoints. Failure of caprock often occurs when dominated by a plunge pool (Fig. 7.3) Streams pouring over headscarps may trigger vertical drilling by falling into plunge pools and abrading owing to the impact of sediments that circulate within the turbid water often with depth of several meters. Plunge pool depth is inversely related to knickpoint retreat (Grimaud et al. 2016). The angle of impact must be steep enough so that a high percentage of the discharge is directed back and underneath the overfall as a reverse roller (May 1989). The turbidity observed within plunge pools suggests that most sediments are in suspension, providing tools for erosion. Well-developed plunge pools indicate a dominant erosive headwall retreat. Undercutting and backwearing by plunge pool erosion at the toe of waterfalls can create a horizontally deep overhang and indentation relative to the lip (Fig. 3.4). Failure of the subcaprock forms a debris-induced-oversteepened-reach beneath waterfalls. The debris is abundant near the front of the knickpoint and become less common farther downstream. Coarse mantle that armors the bedrock and effectively protects it from dissection delays the upslope incisional transfer. Following evacuation of the collapsed material, headwall propagation is to continue.

Experimentally it had been concluded that headward incision produces sediments which move downcurrent and might be often stored temporarily downstream of plunge pools (Fig. 3.4). Here the sediments may accumulate as bars that deflect the flow laterally and widen the channel. Downstream aggradation and burial of former eroded sites and a later incision into the stored sediments make up a *complex response* in which the locus of incision migrates with time upstream and aggradation buries the former degraded sites (Wohl 1998).

7.2 Drainage Area Control

There is a highly significant relationship between the distance of knickpoint recession and the catchment area (Bishop et al. 2005). Knickpoints propagate upstream at a rate that is a power law function of the upstream drainage area (Wobus et al. 2006; Loget and Van Den Driessche 2009), i.e., at an ever-decreasing rate proportional to the gradual decrease in the contributing drainage area and the consequently much smaller stream power. Knickpoints at the same topographic altitudes, in different tributaries, even of variable lithologies, may suggest a similar pulse of incision activated by a similar size of drainage basin.

Based on the conception that the drainage area is the proxy for both sediment and water discharge (Leopold and Miller 1956; Leopold et al. 1964), an upstream knickpoint retreat, proportional to the declining watershed area, causes the knickpoint recession rate to decline through time (Schumm et al. 1987). Modeled knickpoints that migrated quickly when activated by large drainage areas, significantly slowed down when activated by smaller drainage areas. The longer the time duration the lower becomes the migration rate. Experiments by Brush and Wolman (1960) and Leopold et al. (1964) showed the exponential migration curve of knickpoints with the length of the equilibrated reach (Fig. 6.2). The travel distance and the celerity of knickpoints have been suggested as power functions of the drainage area with an exponent between 0.45 and 0.55 (Beckers et al. 2015). High-order drainage systems have a better chance to adjust to base-level drop, whereas lower-order streams will keep steeper (Merritts and Vincent 1989). Knickpoints are more common on first-order streams and their frequency decreases with increase in stream order and power.

The transmission of the base-level fall signal by a propagating knickpoint requires a minimum value of discharge. Small drainage areas are unable to respond to base-level fall. Their limited stream power renders them incapable of substantially incising. A minimum catchment area, which may be just of a few km^2, enables knickpoint retreat and development of a concave profile with its steepest part located below the knickpoins' lip, indicating an active transient status (Castillo et al. 2013). Headward incision will stop when approaching the threshold drainage area needed for incising (Montgomery and Dietrich 1992) which defines the minimum stream power (i.e., a minimum combination of catchment area/discharge and gradient) needed for propagation to proceed. Some reduction of the critical discharge is attained by steepening and narrowing. The two-dimensional form of the drainage area upstream of the knickzone also matters.

7.3 Lag

A certain amount of time is required for morphological trends, related to the base-level, to migrate up-valley. Neither fall nor rise of the base-level affects the upstream fluvial system instantaneously. The distance between the base-level and upstream drainage sites causes the decoupling. As it will take a long time for an incision wave to reach the riverhead, a reach of a stream that has not been influenced yet by the base-level fall will exist upstream of the point reached by incision. River sources may thus not be impacted yet by recent base-level falls.

Similarly, *connection delay* along continental margins is the time needed to connect incipient shelf canyons on a just emerged shelf with a fluvial valley by processes of headward erosion (Van Heijst and Postma 2001). Here too, drainage networks display a certain time lag before reaching their maximum incision efficiency (Kooi and Beaumont 1996). The response to a lowering base-level will be slower and equilibrium will take longer to attain when stream discharge is small

(Bonneau and Snow 1992). A faster profile adjustment is observed in streams with larger drainage basins and discharges (Sinha and Parker 1996).

In streams that erode coarse sediments, the lag between the base-level lowering and the response of the stream is longer than in streams eroded in fine material. Lower sediment concentrations in the flowing channel result in longer lag time of response confirmed across a range of sediment sizes (Bonneau and Snow 1992). Part of the lag may be due to armoring and the time needed to gaining hillslope adjustment to channel incision. Lag may also coincide with the period when bedload is trapped (Cook et al. 2013). Sediment loads that decrease with increasing distance from the outlet contribute to the lag upstream. Retreat from the mainstream junction into a smaller tributary will proceed slower, activated by smaller power supplied by the tributary watershed. Upstream movement of base-level effects will slow down and the lag will lengthen along low order tributaries (Parker 1979).

Incision that propagates upstream, following a base-level fall, can continue for a time even after cessation of the fall, when aggradation starts dominating the lower profile segment (Schumm 1993). The upstream drainage network may be adjusting to former, non-existent base-level conditions, imprisoned in a former history despite different recent setting, i.e., remain in a transient state of adjustment to former sedimentary and geomorphic conditions. The upper and lower segments of a drainage network will thus operate out of phase, demonstrating diachroneity along the river profile (Catuneanu 2006). Experimental results show that the period length of diachroneity depends on the rate of base-level fall. High rates of fall are expected to leave minor connection delays (Van Heijst and Postma 2001). Lag will increase with distance upstream as well as with greater cohesion of the substrate. The greater the drainage area and the higher the rate of the base-level fall, the faster will incision migrate and the shorter will the lag be.

Studying the relationship between active tectonics and fluvial geomorphology leads to address the question of the degree of modification of the base-level control by tectonic disturbances following the interactions between these two external factors that control the drainage system (Maher and Harvey 2008). A river controlled by a migrating nickpoint or knickzone, crossing a young active tectonic area, will demonstrate a complex *base-level tectonic-controlled* evolution. The question of the impact of tectonism on an upstream migrating wave of incision is most relevant. Tectonic movement along faults crossing drainage lines may override trends triggered by base-level aggradation or incision. Faulted river terraces record an initial base-level incisional phase followed by tectonics. Local lowering due to tectonic movement along fault lines may accentuate former base-level incision.

References

Alexandrowicz Z (1994) Geologically controlled waterfall types in the Outer Carpathians. Geomorphology 9(2):155–165

Beckers A, Bovy B, Hallot E, Demoulin A (2015) Controls on knickpoint migration in a drainage network of the moderately uplifted Ardennes Plateau, Western Europe. Earth Surf Proc Land 40(3):357–374
Begin ZB (1978) Aspects of degradation of alluvial streams in response to base-level lowering. Unpublished PhD dissertation, Colorado State University, Fort Collins, Colorado, 239 p
Bishop P, Hoey TB, Jansen JD, Artza IL (2005) Knickpoint recession rate and catchment area: the case of uplifted rivers in Eastern Scotland. Earth Surf Proc Land 30(6):767–778
Bonneau PR, Snow RS (1992) Character of hearwaters adjustment to base level drop, investigated by digital modeling. Geomorphology 5(3–5):475–487
Bressan F, Papanicolaou AN, Abban B (2014) A model for knickpoint migration in first-and second-order streams. Geophys Res Lett 41(14):4987–4996
Brocard GY, Willenbring JK, Miller TE, Scatena FN (2016) Relict landscape resistance to dissection by upstream migrating knickpoints. Journal of Geophysical Research: Earth Surface 121(6): 1182-1203
Brush LM, Gordon Wolman M (1960) Knickpoint behavior in noncohesive material: a laboratory study. Geol Soc Am Bull 71(1):59–74
Bull WB (1991) Geomorphic responses to climatic change. Oxford University Press, Oxford, p 326
Castillo M, Bishop P, Jansen JD (2013) Knickpoint retreat and transient bedrock channel morphology triggered by base-level fall in small bedrock river catchments: the case of the Isle of Jura, Scotland. Geomorphology 180:1–9
Catuneanu O (2006) Principles of sequence stratigraphy. Elsevier, Amsterdam
Cook KL, Turowski JM, Hovius N (2013) A demonstration of the importance of bedload transport for fluvial bedrock erosion and knickpoint propagation. Earth Surf Proc Land 38(7):683–695
Crosby BT, Whipple KX (2006) Knickpoint initiation and distribution within fluvial networks: 236 waterfalls in the Waipaoa River, North Island, New Zealand. Geomorphology 82(1–2):16–38
DiBiase RA, Whipple KX, Lamb MP, Heimsath AM (2015) The role of waterfalls and knickzones in controlling the style and pace of landscape adjustment in the western San Gabriel Mountains, California. Bulletin 127(3–4):539–559
Flores-Cervantes JH, Istanbulluoglu E, Bras RL (2006) Development of gullies on the landscape: a model of headcut retreat resulting from plunge pool erosion. J Geophys Res Earth Surf 111(F1)
Gardner TW (1983) Experimental study of knickpoint and longitudinal profile evolution in cohesive, homogeneous material. Geol Soc Am Bull 94(5):664–672
Grimaud JL, Paola C, Voller V (2016) Experimental migration of knickpoints: influence of style of base-level fall and bed lithology. Earth Surf Dyn 4(1):11–23
Haviv I, Enzel Y, Whipple KX, Zilberman E, Matmon A, Stone J, Fifield KL (2010) Evolution of vertical knickpoints (waterfalls) with resistant caprock: insights from numerical modeling. J Geophys Res Earth Surf 115(F3)
Howard AD, Dietrich WE, Seidl MA (1994) Modeling fluvial erosion on regional to continental scales. J Geophys Res Solid Earth 99(B7):13971–13986
Kooi H, Beaumont C (1996) Large-scale geomorphology: classical concepts reconciled and integrated with contemporary ideas via a surface processes model. J Geophys Res Solid Earth 101(B2):3361–3386
Lamb MP, Howard AD, Dietrich WE, Perron JT (2007) Formation of amphitheater-headed valleys by waterfall erosion after large-scale slumping on Hawaii. Geol Soc Am Bull 119(7–8):805–822
Leopold LB, Miller JP (1956) Ephemeral streams: Hydraulic factors and their relation to the drainage net. US Government Printing Office
Leopold LB, Wolman MG, Miller JP (1964) Fluvial processes in geomorphology. Hydraulic factors and their relation to the drainage net. US. WH Freeman and Company, San Francisco, p 522
Loget N, Van Den Driessche J (2009) Wave train model for knickpoint migration. Geomorphology 106(3–4):376–382
Maher E, Harvey AM (2008) Fluvial system response to tectonically induced base-level change during the late-Quaternary: the Rio Alias southeast Spain. Geomorphology 100(1–2):180–192

May JH (1989) Repair, evaluation, maintenance and rehabilitation research program. Geotechnical aspects of rock erosion in emergency spillway channels. Report 4: Geologic and hydrodynamic controls on the mechanics of knickpoint migration. Army Engineer Waterways Experiment Station Vicksburg Ms Geotechnical Lab

Merritts D, Vincent KR (1989) Geomorphic response of coastal streams to low, intermediate, and high rates of uplift, Medocino tripple junction region, northern California. Geol Soc Am Bull 101(11):1373–1388

Montgomery DR, Dietrich WE (1992) Channel initation and the problem of landscape scale. Science 255:826–830

Parker G (1979) Hydraulic geometry of active gravel rivers. J Hydraul Div 105(9):1185–1201

Robl J, Heberer B, Prasicek G, Neubauer F, Hergarten S (2017) The topography of a continental indenter: the interplay between crustal deformation, erosion, and base-level changes in the eastern Southern Alps. J Geophys Res Earth Surf 122(1):310–334

Schumm SA (1993) River response to base-level change: implications for sequence stratigraphy. J Geol 101(2):279–294

Schumm SA, Harvey MD, Watson CC (1984) Incised channels. Water Resources Publications, 200 p

Schumm SA, Mosley MP, Weaver W (1987) Experimental fluvial geomorphology

Sinha SK, Parker G (1996) Causes of concavity in longitudinal profiles of rivers. Water Resour Res 32(5):1417–1428

Sklar LS, Dietrich WE (2004) A mechanistic model for river incision into bedrock by saltating bed load. Water Resour Res 40(6)

Van Heijst MW, Postma G (2001) Fluvial response to sea-level changes: a quantitative analogue, experimental approach. Basin Res 13(3):269–292

Whipple KX, Kirby E, Brocklehurst SH (1999) Geomorphic limits to climate-induced increases in topographic relief. Nature 401(6748):39–43

Wobus CW (2005) Geomorphic and thermochronologic signatures of active tectonics in the central Nepalese Himalaya. Doctoral dissertation, Massachusetts Institute of Technology

Wobus C, Whipple KX, Kirby E, Snyder N, Johnson J, Spyropolou K, Crosby B, Sheehan D (2006) Tectonics from topography: procedures, promise, and pitfalls. Special papers—Geological Society of America 398, Penrose conference series, pp 55–74

Wohl EE (1998) Bedrock channel morphology in relation to erosional processes. In: Geophysical monograph. American Geophysical Union 107, pp 133–152

Knickpoint Evolution

8

Abstract

While propagating upstream, knickpoints do not retain their form. *Form evolution* during headward knickpoint migration may include diffusion, parallel retreat, rotation or replacement. The propagating wave of erosion shows reduction in erosion rate up the drainage basin. The effect of a lowering base-level often propagates only a relatively limited distance upstream and then diffuses. Only a combination of base-level lowering plus a tectonic inland uplift will increase the total base-level fall impact, and activate the knickpoint migration and evolution to the far inland drainage systems.

Keywords

Diffusion • Transient boundary • Knickpoint partition • Static knickpoint • Parallel retreat • Knickpoint rotation

8.1 Knickpoint Diffusion

Based on experiments (Grimaud et al. 2016), new knickpoints are generated each time the base-level drops. In between the drops, during quiescent periods, the profile can smoothly readjust by overall *diffusion*, i.e., during upstream migration, becoming gradually smooth by changing the gradient and shape (Figs. 8.1 and 8.2). Diffusion is linearly correlated with the rain index, i.e., the mean annual rain volume over a basin as a proxy for discharge (Ben-Moshe et al. 2008). The diffusivity coefficient $K = \text{m}^2\text{year}^{-1}$ predicts the smoothing through time while scaling positively with sediment discharge and negatively with the channel width (Daniels 2008). Values of K derived from published sources vary for natural channels approximately 102–108 m^2 year^{-1} (Castelltort and Van Den Driessche 2003; Doyle et al. 2003). A mathematical model that describes the channel response to base-level lowering and utilizes a diffusion equation has been successfully applied following flume experiments by Begin et al. (1980, 1981) and Begin (1988).

D. Bowman, *Base-level Impact*,
https://doi.org/10.1007/978-3-031-24994-5_8

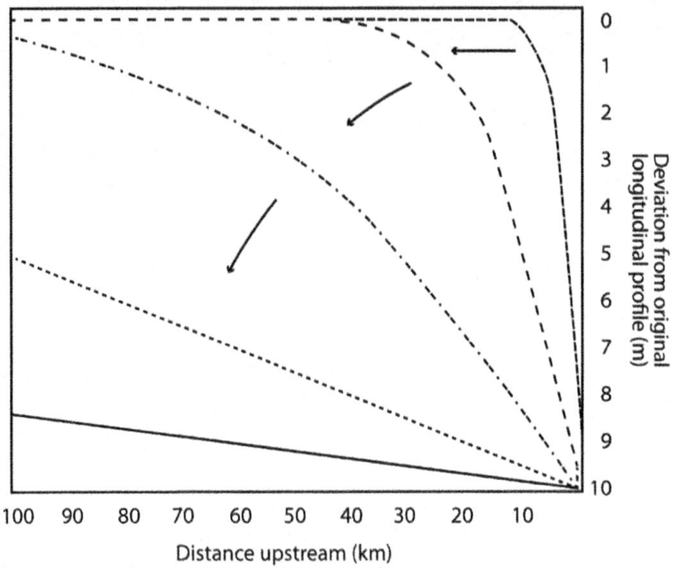

Fig. 8.1 Longitudinal profile development by diffusion. Defined are the vertical bed elevation change and the horizontal migration of the bed disturbance. Drop of 10 m at the channel outlet (sharp profile change at right) results in propagation of the knickpoint upstream with a gradual smoothing of the longitudinal gradient by lowering of the channel bed elevation by the flows. The upstream boundary that limits the diffusive incision is not shown (Redrawn from Daniels 2008)

If no significant diffusion takes place, knickpoints may start to migrate and retreat upstream as a leading edge of an inland advancing escarpment or step, demarcating the upstream end of the base-level fall control, i.e., a transient boundary between the upstream higher channel and the adjusting, new, lower, incised segment. Upstream migration of knickpoints and establishing steeper hillslopes is a primary mechanism of landscape adjustment to base-level changes (Harkins et al. 2007). Along the profile *each knickpoint functions as a local base-level.* The network upstream of the knickpoint is not affected by the terminal base-level.

In a layered substrate of contrasting strength, *Knickpoint partitioning* may result by splitting into discrete knickpoints that propagate upstream at different rates. The transient signal becomes divided into two parts (Fig. 8.3), i.e., the lower knickpoint may remain localized on the lithologic boundary, while the upper knickpoint sweeps rapidly through the softer bedrock (Cook et al. 2009). Non-cohesive non-layered sediments that experience a similar base-level lowering are more likely to form a non-stepped knickpoint.

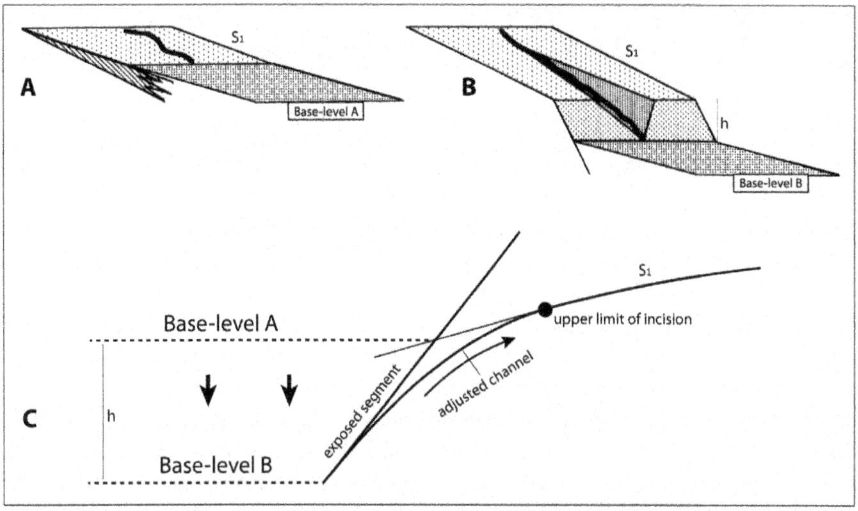

Fig. 8.2 Upstream knickpoint diffusion following base-level fall: **A** Channel crossing a plain of a uniform gentle slope (S1). **B** Following tectonic subsidence the base-level falls, exposing an over-steepened segment (*h*) that erodes headwards. **C** The eroded segment does not migrate upstream through the entire network but diffuses at a certain distance where it adjusts to the initial slope, schematic (Redrawn following Salter 2016)

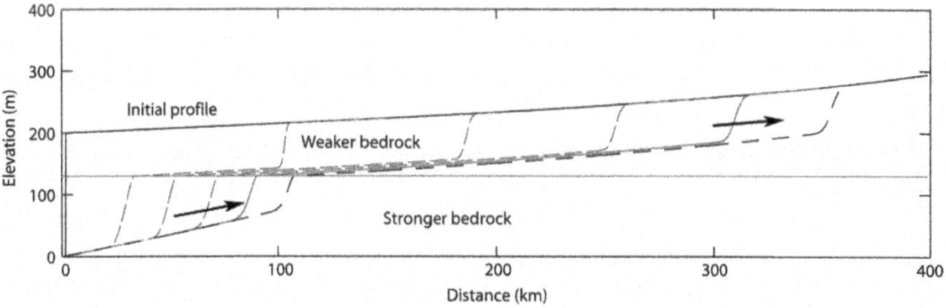

Fig. 8.3 Evolution of a profile incising through a two-layer bedrock in response to base-level fall. The upper, weaker and more erodible unit sweeps quicker through the system forming a lithologic partitioning knickpoint. Based on a numerical model (Modified after Cook et al. 2009)

Apart from evolving as mobile features that migrate upstream, knickpoints may as well form *anchored features* and become static due to the presence of a resistant lithology or may decrease their migration velocity and steepen. Such evolution is expected when the channel is nourished by a small drainage area and is below the threshold of transport capability which leads to stagnation (Crosby and Whipple 2006). Such evolution generates "*hanging* valleys" as a consequence of the inability of the tributary to maintain in pace with the trunk stream.

8.2 Upstream Propagation

How far inland does the base-level control the morphology (Schumm 1993)? Base-levels control only a certain distance upstream and their effect often diffuses with different amounts (Bonneau and Snow 1992). It would be most unlikely for a knickpoint to travel the entire length to the headwater (Brush and Wolman 1960). In unconsolidated sediments with channels capable of transporting the material easily, headward knickpoint migration is commonly limited. Knickpoints will not move upstream indefinitely. There is a distance where the knickpoint profile diminishes and the channels do not respond anymore to the base-level control (Fig. 8.2). In homogeneous unconsolidated material, knickpoints tend to migrate upstream for only a short distance before becoming too faint to recognize. Cohesive sediments can help a propagating knickpoint maintain its shape, not to diffuse with distance and progress farther upstream, leading to increased headward incision. Maximum distance a knickpoint may travel, given unlimited time, may be controlled by the ratio of the slope of the oversteepened reach to the average slope of the channel. The greater the ratio of the slope of the knickpoint to the average slope of the channel, the farther upstream will a knickpoint migrate (Brush and Wolman 1960).

Knickpoints will almost stop moving and become stabilized if the reduction in slope minimizes their competence to move material or when controlled by a resistant outcrop (Brush and Wolman 1960). Base-level-induced incision may propagate up an alluvial fan by distances of up to several kilometers over timescales of hundreds of years. Propagation of the effects of a base-level change upstream within a range of tens of kilometers may take tens to hundred thousands of years. The up-system propagation period of a base-level change may be longer than the timescales of Quaternary climatic changes (Harvey 2002).

A large and deep entrenchment that extends far from the coast may require an unrealistic base-level fall, thus justifying the assumption of an uplift, which may cause significant steepening, or a change in the water/sediment regime that can enhance the incision rate (Schumm 1993). Only a combination of base-level lowering, which rejuvenates the lower part of the fluvial system, plus tectonic uplift that rejuvenates the upper part of the drainage network may activate most of the drainage system (Catuneanu 2006). A large-scale effect on the distance of the base-level control can also be, in the long term, river capture and drainage re-routing.

The influence of the late-Wisconsin sea level fall extended up valley about 100 km in case of the Colorado River in Texas and to more than 200 km in the case of other larger rivers. The influence of the 120 m sea level fall on the Mississippi extended 370 km upstream out of more than 1500 km (Shanley and McCabe 1993). The sub-Dakota unconformity in the southeastern Colorado Plateau suggests that eustasy may have played an important role even hundreds of kms well inland from the sea (Aubrey 1989).

The headwater is the domain of the smaller streams with lower sediment fluxes and lower discharge, forming top-down processes that are less capable of incising channel beds and lowering hillslopes and are thus *outside* the significant base-level impact. The headwaters steepen to supply sufficient stream power so that erosion can continue (Seidl et al. 1992) but do not pass from the transient landscape stage to full steady state (Bishop 2007).

8.3 Form Evolution

Laboratory flume experiments on channel response to instantaneous base-level fall showed that knickpoints do not retain their form as they propagate upstream (Gardner 1983; Holland and Pickup 1976). Different scenarios of knickpoint evolution and migration in homogeneous, horizontally bedded substrates of variable resistance have been summarized by Brush and Wolman (1960), Frankel et al. (2007) and by Leopold et al. (1964).

The evolution of the form of a knickpoint depends on the relative magnitudes of the incision vectors (Fig. 8.4) and to a great degree on the rate of erosion at the base, relative to that on the lip (Whipple et al. 2013). The incision vectors include: (1) vertical incision above the knickpoint, (2) horizontal erosion of the knickpoint face and (3) vertical incision at the base of the knickpoint. *Parallel retreat* with a vertical or near-vertical knickpoint face retreat will occur when the highest rates of erosion are focused at the face and base, undercutting and steepening the knickpoint. Undercutting processes dominate especially where there is a strong-over-weak stratigraphy with the ability of the flow to evacuate from the base material that had collapsed, allowing the knickpoint to maintain a steep face and keep its propagation rate. Parallel retreat is the most common scenario in layered rocks of differing erosional resistance (Gilbert 1896; Gardner 1983).

Form evolution can be explained aswell based on $\tau_c - \tau_0$ *relations,* when τ_c the critical shear stress needed to initiate erosion and τ_0 the actual shear stress (Fig. 8.5). *Knickpoint replacement* will occur when the head of the knickpoint moves faster than its base so that it "rolls back" as it migrates upstream (Gardner 1983). Thus the knickpoint tends to lengthen, flatten, gradually looses its identity by reclining and becoming nearly equal to the stream (Holland and Pickup 1976) by finally grading into the upper unincised surface and becoming indistinct (Begin 1978). The decrease in the slope of the knickpoint, as it propagates upstream to the point where it cannot be distinguished, demonstrates a negative feedback loop promoting decay. Knickpoints may die out and diffuse in the vicinity of the base-level fall without prolonged distance retreat (Tucker and Slingerland 1994).

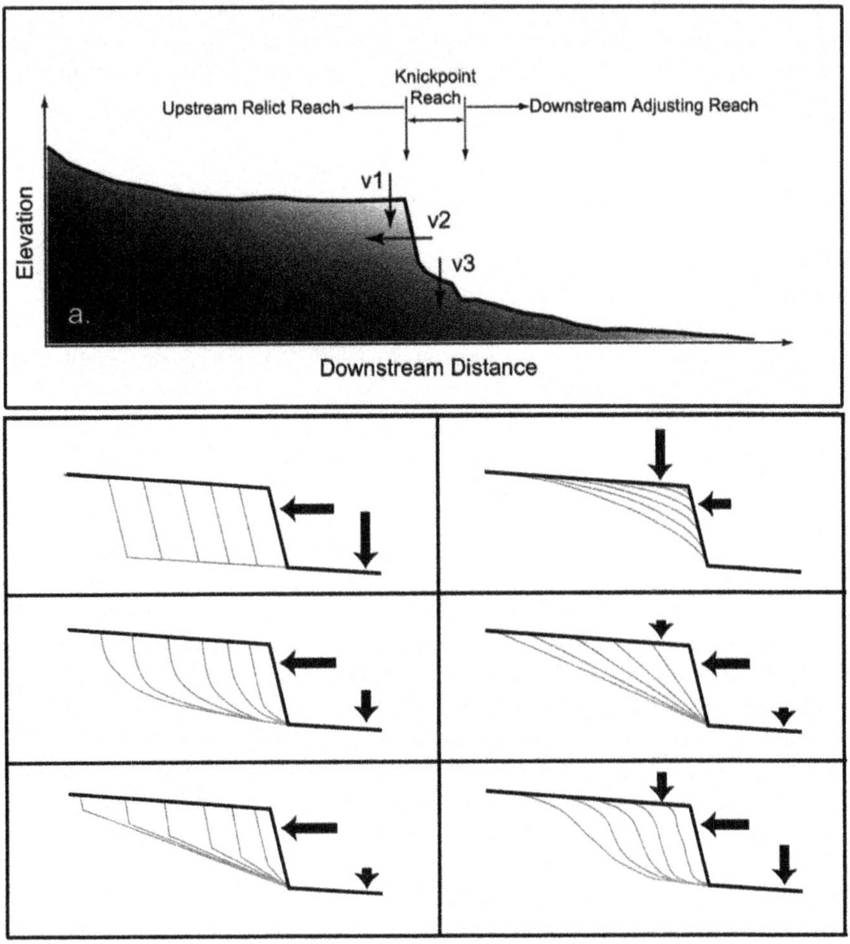

Fig. 8.4 Evolution of the knickpoint form when dependent on relative magnitudes of incision vectors. Different combinations of the vectors' magnitudes result in diverse final knickpoint morphologies. Knickpoint flattening is analogous to diffusion. Schematic (From Crosby and Whipple 2006)

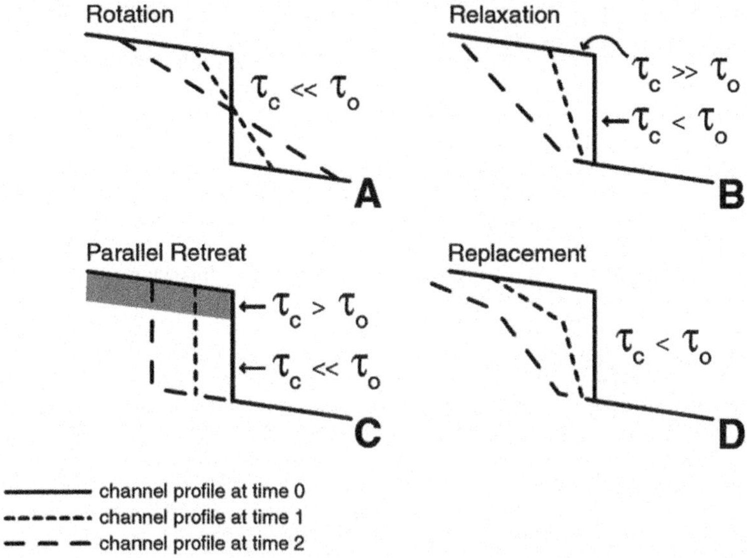

Fig. 8.5 Models of a waterfall knickpoint retreat and form evolution based on $\tau_c - \tau_0$ relations. τ_c—the critical shear stress needed to initiate erosion, τ_0—the actual shear stress. **A** Uniformly nonresistant material shows face inclination with downstream aggradation. **B** Uniformly very resistant material demonstrates face inclination without aggradation. **C** Resistant above nonresistant layers show parallel retreat by undermining. **D** Uniformly moderately resistant, Schematic (From Gardner 1983)

References

Aubrey WM (1989) Mid-cretaceous alluvial-plain incision related to eustasy Southeastern Colorado Plateau. Geol Soc Am Bull 101:44

Begin ZB (1978) Aspects of degradation of alluvial streams in response to base-level lowering. Unpublished Ph.D. dissertation, Colorado State University Fort Collins, Col, 239p

Begin ZB (1988) Application of diffusion-erosion model to alluvial channels which degrade due to base-level lowering. Earth Surf Proc Land 13:487–500

Begin ZB, Meyer DF, Schumm SA (1981) Development of longitudinal profiles of alluvial channels in response to base-level lowering. Earth Surf Proc Land 6(1):49–68

Begin ZB, Schumm SA, Meyer DF (1980) Knickpoint migration due to base-level lowering. J Waterway Port Coastal Ocean Div 106(3):369–388

Ben-Moshe L, Haviv I, Enzel Y, Zilberman E, Matmon A (2008) Incision of alluvial channels in response to a continuous base-level fall: field characterization, modeling and validation along the Dead Sea. Geomorphology 93:524–536

Bishop P (2007) Long-term landscape evolution: linking tectonics and surface processes. Earth Surf Proc Land 32(3):329–365

Bonneau PR, Snow RS (1992) Character of headwaters adjustment to base-level drop, investigated by digital modeling. Geomorphology 5(3–5):475–487

Brush LM Jr, Gordon Wolman M (1960) Knickpoint behavior in noncohesive material: a laboratory study. Geol Soc Am Bull 71(1):59–74

Castelltort S, Van Den Driessche J (2003) How plausible are high-frequency sediment supply-driven cycles in the stratigraphic record? Sed Geol 157(1–2):3–13

Catuneanu O (2006) Principles of sequence stratigraphy. Elsevier, Amsterdam

Cook KL, Whipple KX, Heimsath AM, Hanks TC (2009) Rapid incision of the Colorado River in Glen Canyon—insights from channel profiles, local incision rates, and modeling of lithologic controls. Earth Surf Proc Land 34(7):994–1010

Crosby BT, Whipple KX (2006) Knickpoint initiation and distribution within fluvial networks: 236 waterfalls in the Waipaoa River, North Island, New Zealand. Geomorphology 82(1–2):16–38

Daniels JM (2008) Distinguishing allogenic from autogenic causes of bed elevation change in late quaternary alluvial stratigraphic records. Geomorphology 101(1–2):159–171

Doyle MW, Stanley EH, Harbor JM (2003) Channel adjustments following two dam removals in Wisconsin. Water Resources Res 39

Frankel KL, Pazzaglia FJ, Vaughn JD (2007) Knickpoint evolution in a vertically bedded substrate, upstream-dipping terraces and Atlantic slope bedrock channels. Geol Soc Am Bull 119(3–4):476–486

Gardner TW (1983) Experimental study of knickpoint and longitudinal profile evolution in cohesive, homogeneous material. Geol Soc Am Bull 94(5):664–672

Gilbert GK (1896) The underground water of the Arkansas Valley in eastern Colorado. US Government Printing Office

Grimaud JL, Paola C, Voller V (2016) Experimental migration of knickpoints: influence of style of base-level fall and bed lithology. Earth Surf Dyn 4(1):11–23

Harkins N, Kirby E, Heimsath A, Robinson R, Reiser U (2007) Transient fluvial incision in the headwaters of the Yellow River, Northeastern Tibet, China. J Geophys Res: Earth Sur 112(F3)

Harvey AM (2002) Effective timescales of coupling within fluvial systems. Geomorphology 44(3–4):175–201

Holland WN, Pickup G (1976) Flume study of knickpoint development in stratified sediment. Bull Geol Soc Am 8:76–82

Leopold LB, Wolman MG, Miller JP (1964) Fluvial Processes in geomorphology. W.H. Freeman and Co, San Francisco, 522p

Salter T 2016 Fluvial scour and incision: models for their influence on the development of realistic reservoir geometries. Geological Society of America, Special publication

Schumm SA (1993) River response to base-level change: implications for sequence stratigraphy. J Geol 101(2):279–294

Seidl MA, Dietrich WE, Schmidt KH, de Ploey J (1992) The problem of channel erosion into bedrock. Functional Geomorphol 101–124

Shanley KW, Mc Cabe PJ (1993) An aid to sequence stratigraphic interpretation of alluvial strata. In: Proceedings of the 7th IFP exploration and production research conference held in Scarborough

Tucker GE, Slingerland RL (1994) Erosional dynamics, flexural isostasy, and long-lived escarpments: a numerical modeling study. J Geophys Res: Solid Earth 99(B6):12229–12243

Whipple KX, Dibiase RA, Crosby BT (2013) Bedrock rivers. In: Treatise on geomorphology. Elsevier, Amsterdam, pp 550–573

Longitudinal Profiles

9

Abstract

Longitudinal profiles may provide clues to the impact of base-levels, to active erosional processes and tectonic uplift or subsidence. Graded profiles approximate a smooth, concave-upward continuous curve that has been claimed to be a necessary marker for steady state. Graded profiles are controlled by a long-term balance between sediment supply and sediment transport capacity. Steep channels at the upslope end of drainage basins develop self-enhancing feedback conditions for incision. Channel convexities appear to indicate a lithologic, tectonic or a transient erosional front. When crossing lithologic boundaries, longitudinal profiles may become completely unrelated to the base-level impact. The profiles always adjust to the first adjacent downstream local base-level, not to the ultimate one.

Keywords

Lithological knickpoint • Concavity • Transient response • Divergence • Steady-state profile • Local base-level

The longitudinal profile is a geometrical description of the channel and serves as an efficient primary tool in base-level studies. Streams shape their longitudinal profile by cutting or filling in order to continue and carry the sediment load with their given water discharge (Lane 1955). Longitudinal profiles provide clues to the impact of the base-level (Fig. 9.1), reflect the underlying materials, uncover active erosional processes and possible tectonic movements, all which constitute the basic geomorphological history.

The altitude, shape and gradient of longitudinal profiles are described in terms of *steepness, concavity* and *convexity* (Hack 1957; Flint 1974). Varying lithologies of bedrock usually form stepped profiles with repetitive knickpoints. The profile of streams that plot nearly straight lines on a semi-log graph indicates adjustment to the material over which they flow. Departure from linearity may reflect adjustment

D. Bowman, *Base-level Impact*,
https://doi.org/10.1007/978-3-031-24994-5_9

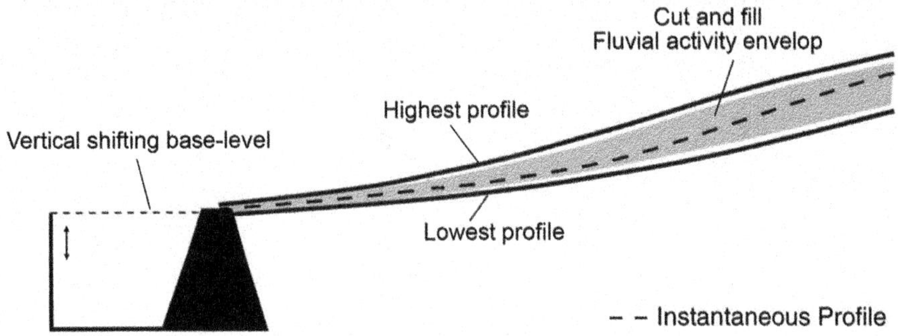

Fig. 9.1 Vertical envelope of a profile activity: the longitudinal profile of a channel is anchored to its base-level which controls the profile by shifts of elevation that trigger cut-and-fill cycles. The entire channel activity is bounded vertically by the highest possible aggradation and the maximum possible incision below. The uppermost and lowest profiles mark an envelope which defines the vertical space of the profile activity through time. An instantaneous profile indicates a single point in time within the evolution of the profile (redrawn from Holbrook et al. 2006)

to a different control and indicates disequilibrium (Hack 1973; Reed 1981). Knickpoints compose discontinuities in the profile. A lithological knickpoint makes a steep channel reach that separates between two somewhat more moderate channel segments downstream and upstream (Fig. 3.1).

9.1 Concavity, Convexity, Divergence and Convergence

The channel slope typically follows a power law relationship with the upstream drainage area. Slope declines with increasing drainage area, resulting in a *concave-up* profile form. Following theory, alluvial rivers increase in width and decrease in slope downstream, offsetting the increasing discharge and maintaining a somewhat consistent distribution of stream power and minimum geomorphic work along the concave profiles (Leopold and Langbein 1962).

The altitude and the proximity of base-levels to the divide control the overall steepness of the profiles. The concave shape manifests a downstream increase in discharge and controls the rate at which potential energy is dissipated. Steady-state profiles are dominated by a transition from a coarse, weakly mobile bed with extensive bedrock exposure in the upstream reaches to a finer, highly mobile bed with limited bedrock exposures downstream. Concave-upward profiles are steep close to their source and flatten with distance downstream as the base-level is approached. Lack of knickpoints along a profile suggests that the channel is responding to its terminal base-level in a continuous way (Cook et al. 2009). Long duration of activity and relative soft rocks contribute toward less knickpoints.

Deviation from concavity indicates immaturity. Concavity has been claimed by many researchers to be a necessary (but not sufficient) indicator for steady state (Chatanantavet and Parker 2009). However, fully relaxed concave-up profiles are often not to be expected. Several processes can cause a longitudinal river profile to deviate from simple concavity and form a local convex-up segment. These include formation of a landslide dam and crossing a lithologic contact or an uplifted rock.

For steady state to occur, the rate of rock uplift must be balanced by the rate of bedrock incision along the profile. At any given point along the profile, sufficient shear stress must be available to erode the bedrock at the rate of rock uplift (Sklar and Dietrich 2008) and mobilize the coarse sediment from upstream. When incision and uplift rates balance each other along the profile, concave profiles will evolve, marking an equilibrium state. The concave channel slope is most sensitive to changes in grain size delivered to the channel from the hillslopes. During the initial response to base-level fall, changes in sediment flux activated by the hillslope response lag behind the profile adjustment.

The history of tectonic uplift (base-level lowering) can be well analyzed using longitudinal profiles as dynamic geomorphic markers (Hilley and Arrowsmith 2008). Low-order channels at the upstream, due to their relative small drainage area, are unable to incise to an amount controlled by the base-level fall and will become with time the steepest segments in the network. Steepening upstream leads to reduced channel and valley width and allows more concavity to be regained. At the upslope end of drainage basins, the steeper channels form strong, self-enhancing, feedback conditions for incision (Goode and Wohl 2010). When headwater segments of channels are denudated and finally lowered, channels become less concave. Length added to channels downstream decreases the concavity. The decrease in slope at the lower course of concave rivers, toward the base-level, strongly influences bedload transport capacity and cause reduction in competence (Catuneanu et al. 2011), enhancing a negative feedback by decreasing stream power and the incision rate.

Profiles often show a *convex section*, i.e., a segment where the channel gradient gradually increases downstream (Fig. 9.2). The convex step separates the lower adjusting profile reach from the upper channel segment that has not responded yet to the base-level fall. Convexities, whether knickpoints or knickzones, can thus serve as morphological markers for unravelling tectonics or lithologic hardening. If a tectonic uplift (falling base-level) or lithological variability are both of no potential explanation and no other explanations concerning the profile convexity exists, we may infer that a transient signal of a falling base-level has reached the site (Fig. 3.3).

Numerous investigators who studied the fluviatile response to a lowering base-level reported on convex profiles: Holland and Pickup (1976) reported on an experiment of a stream that cut through a coastal escarpment and formed a convex profile. Pickup (1977) showed convexity in a simulation model of back cutting entrenchment. Convex longitudinal profiles resulting from knickpoint replacement in uniform, moderately resistant material, or through a more resistant lithology, were reported by Begin et al. (1981), observed in a flume. This experiment suggested a convex longitudinal profile development from the initial degradational stage up to the accommodation to the lowered base-level, implying convexity to dominate during the entire transient phase (Fig. 9.3). Wolman (1987) reported of

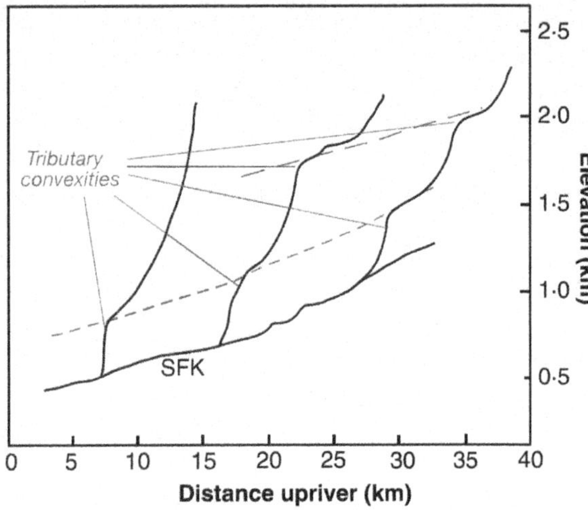

Fig. 9.2 Longitudinal profiles of tributary streams entering the South Fork Kings (SFK) inner gorge, the Sierra Nevada, California. The tributaries show convex knickzones as they cascade steeply into an inner gorge (modified after Stock et al. 2005)

convex profiles during the intermediate stage of headcut progress through weathered saprolite. Seidl et al. (1994) reported on stream channels with graded to slightly convex profiles on the Hawaii Islands as a result of boulder deposition by landslides. The boulders armored the bed and slowed down entrenchment. Clear convexities were observed as well along the Truckee River in western Nevada, incised in response to a 20 m lowering of Pyramid Lake over the past 120 years, leaving a suite of cut terraces that extend ~ 15 km upstream from the lake. From the geometry of reconstructed stream profiles, based on the terraces remnants, it was possible to infer the history of the base-level fall (Adams 2012).

9.2 Controls of the Profile

Discrepancy between reconstructed profiles may display the history of the entrenchment and the base-level fall. Downstream diverging profiles document control by a fast falling base-level (Fig. 9.4). Parallel concave profiles represent continuous entrenchment with periods of steady state that allow identical profile adjustment. Downstream converging profiles mark a stable base-level where to the profiles gradually grade. The following are the main factors controlling the shape of longitudinal profiles:

Lithology—Changes in longitudinal profiles are often closely related to differences in lithology. When a hard layer is exposed, erosion rates decrease and slopes increase. Resistant lithologies have the effect of steepening the profile and exert a first-order control on the rate of knickpoints' migration. Lithologic contrasts can produce differences in channel steepness. Very slow migrating lithologic knickpoints may become almost permanent anchored features. Profile evolution may thus reflect *both* the regional heterogeneous lithology *and* the base-level impact, both generating a *complex evolution history* (Yanites et al. 2017).

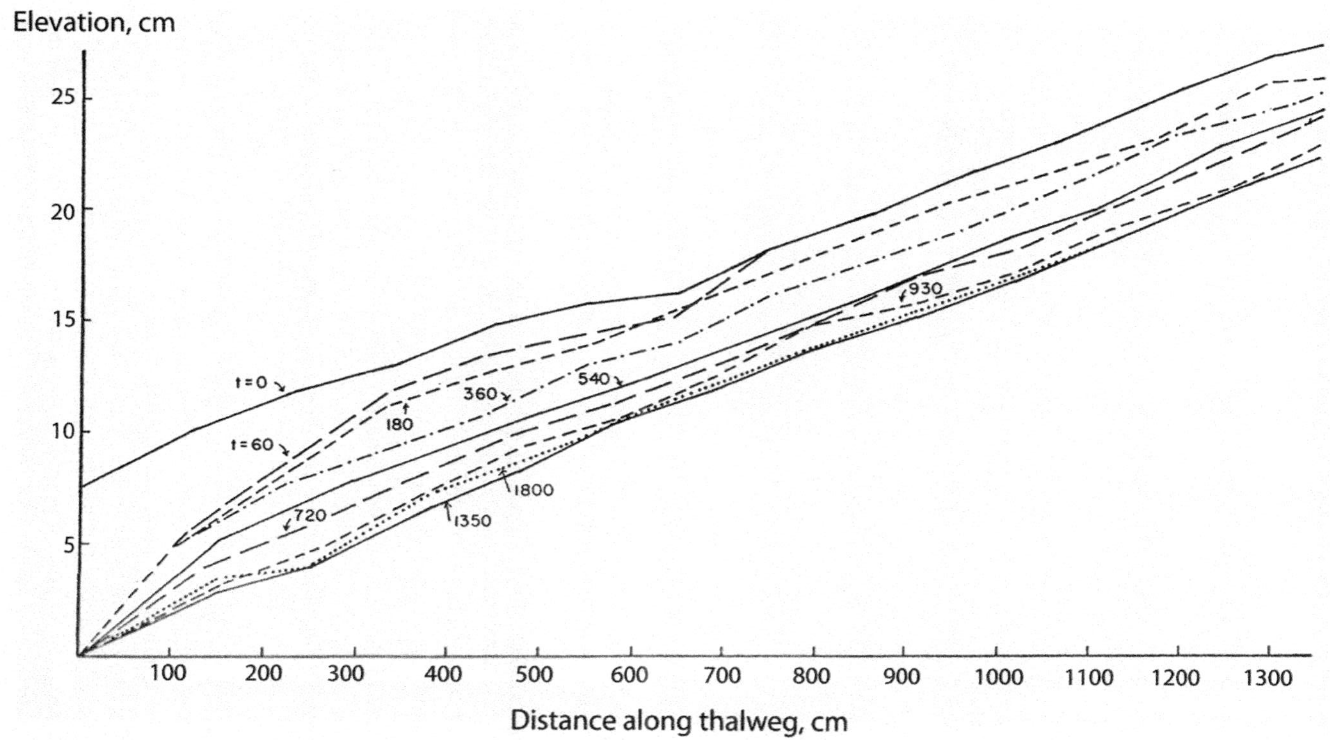

Fig. 9.3 Experimental longitudinal profile development following base-level fall. Note the initial convexity that makes a diverging profile pattern when related to the upper surface and a converging pattern to the stable experimental base-level. Demonstrated is diffusion as well. The new profile induced by the base-level fall shows essentially the same form as the original profile (from Begin et al. 1981)

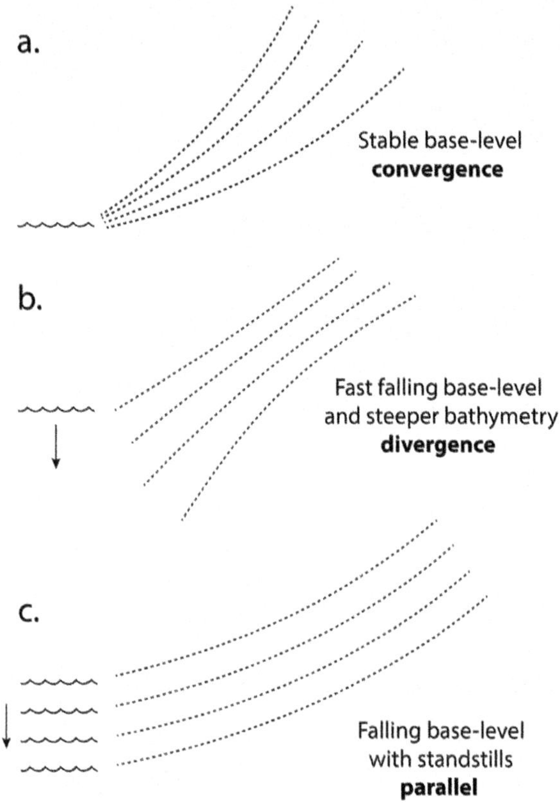

Fig. 9.4 Paleo-profiles represented by river terraces outline the base-level dynamics by divergence, convergence or parallelism. Converging terraces **a** mark base-level stability. Diverging profiles **b** result of a rapid base-level fall. The profiles try to trace the subsiding base-level and may also be due to spatial variation in channel incision capacity. Parallel profiles **c** indicate profiles in a similar standstill and adjustment position during a continuous base-level fall

Weaker rocks adjust their slopes to a base-level fall rate more rapidly than harder rocks (Fig. 8.3). When multiple rock types are present, a transient erosion wave tends to travel at different rates through the different erodibilities and accentuate or limit the migration of the base-level signal.

Sediment supply to the channel over time is of first-order importance in controlling the profile evolution. Incision rate is enhanced by availability of abrasive tools and limited by bedrock exposure. Veneers of coarser sediment contribute to armoring and maintenance of the bed by preventing degradation. Lower slopes of profiles with a coarser sediment veneer may indicate detached-limited conditions with incomplete adjustment to the base-level. Following weathering, abrasion and breakdown of boulders, evacuation of the coarse lag veneer becomes possible. Material will be removed by undermining and boulder rollover triggered by rare catastrophic floods. The mobility of the coarse grains is expected to be more sensitive to base-level changes than the movement of fines (Chavarrías et al. 2018). A significant fraction of debris may often still remain non-transportable and thus steepen the channels and lower the average knickpoint retreat rate (Haviv et al. 2010).

North–south longitudinal profile-aligned rivers of the Bhutan Himalaya Mountain Range show irregular profiles with steep concave sections separated by convex knickzones (Baillie and Norbu 2004). The base of the concave sections is suggested to indicate a paleo-base-level. Concave sections that are separated by convex knickzones, if not related to lithology, are probably result from hiatuses in the continuity of tectonic uplifts (base-level falls), i.e., indicating periods of a relative stable base-level (Stock et al. 2004; Baillie et al. 2004). A regional repetition of convex sections succeeding concave segments, at similar altitudes without identical lithology, could suggest a paleo-regional base-level control.

Contributing drainage area—Streams commonly have power-law scaling between the channel slope S and the upstream contributing drainage area A. This empirical relationship

$$S = \mathrm{KS}\, A^{-\theta}$$

includes the steepness index KS and the concavity index θ that falls in the narrow range 0.3–0.6 (Wobus et al. 2006). The geomorphic transport laws predict inverse relationship between local slope and drainage area (Seidl et al. 1992). The larger the drainage area, the smaller the slope of the incising profile, while it is still above the threshold line of the shear stress needed for incision (Begin and Schumm 1979). However, channels are steeper and narrower for the same drainage area when crossing faster slipping fault blocks with higher throw rates (faster falling base-level).

Analyzing a region in terms of its longitudinal profiles, while recognizing knickpoints and knickzones, may be a valuable approach to start and infer controlling factors (Fig. 9.5) and phases of incision controlled by base-level changes. The depth of incision, following base-level fall, can be estimated by reconstructing a stream profile, projecting it downstream and subtracting the modern active profile elevation (Fig. 3.6). Due to subsequent deformation of profiles in tectonic active areas, it may be hard or impossible to separate the original profile from a deformed one (Finnegan 2013).

Each knickpoint along a profile *functions as a local base-level* that controls the longitudinal upstream segment. Rivers always adjust to the ***first*** adjacent downstream local base-level (Mackin 1948). Streams react to this highest, closely adjoining local base-level and ***not*** to the far removed, ultimate base-level downstream (Leopold and Bull 1979). Local base-levels (i.e., knickpoints) along a stream contribute to segmentation of the profile, i.e., transforming the profile into a sequence of locally graded, welded segments. The channels thus become separated into reaches of distinct steepness *unrelated to the ultimate terminal base-level* far downstream. Such irregularity along the profile suggests that *the control of the ultimate base-level is often not relevant.*

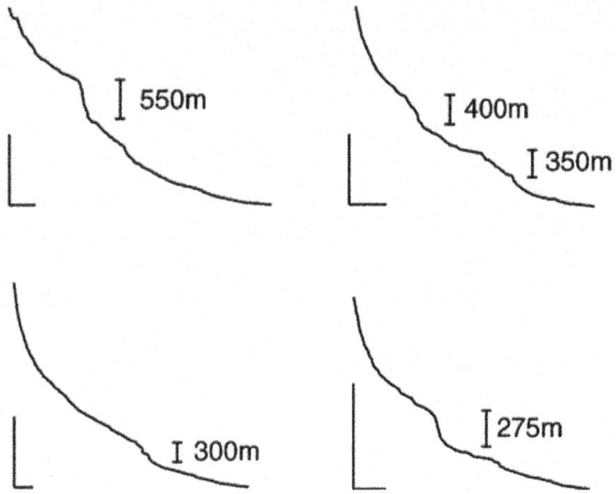

Fig. 9.5 Examples of river profiles in Taiwan that do not follow the power-law slope–area relationship and are punctuated by abrupt changes in channel slope that form knickzones. The knickzones range in relief from 250 to 600 m and in length from 0.6 to 4.5 km and often, but not always, occur at transitions of lithological units. Horizontal and vertical scales are 5 km and 1 km, respectively (from Stolar et al. 2007)

References

Adams KD (2012) Response of the Truckee River to lowering base level at Pyramid Lake, Nevada, based on historical air photos and LiDAR data. Geosphere 8(3):607–627

Baillie IC, Norbu C (2004) Climate and other factors in the development of river and interfluve profiles in Bhutan, Eastern Himalayas. J Asian Earth Sci 22(5):539–553

Begin ZB, Meyer DF, Schumm SA (1981) Development of longitudinal profiles of alluvial channels in response to base-level lowering. Earth Surf Proc Land 6:49–68

Begin ZB, Schumm SA (1979) Instability of alluvial valley floors: a method for its assessment. Am Soc Agric Eng Trans 22:347–350

Catuneanu O, Galloway WE, Kendall CGSC, Miall AD, Posamentier HW, Strasser A, Tucker ME (2011) Sequence stratigraphy: methodology and nomenclature. Newsl Stratigr 44(3):173–245

Chatanantavet P, Parker G (2009) Physically based modeling of bedrock incision by abrasion, plucking, and macro abrasion. J Geophys Res Earth Surf 114(F4)

Chavarrías V, Blom A, Orrú C, Martín-Vide JP, Viparelli E (2018) A sand-gravel Gilbert delta subject to base-level change. J Geophys Res Earth Surf 123(5):1160–1179

Cook KL, Whipple KX, Heimsath AM, Hanks TC (2009) Rapid incision of the Colorado River in Glen Canyon–insights from channel profiles, local incision rates, and modeling of lithologic controls. Earth Surf Proc Land 34(7):994–1010

Finnegan NJ (2013) Interpretation and downstream correlation of bedrock river terrace treads created from propagating knickpoints. J Geophys Res Earth Surf 118(1):54–64

Flint JJ (1974) Stream gradient as a function of order, magnitude, and discharge. Water Resour Res 10(5):969–973

Goode JR, Wohl E (2010) Substrate controls on the longitudinal profile of bedrock channels: implications for reach-scale roughness. J Geophys Res Earth Surf 115:F3

Hack JT (1957) Studies of longitudinal stream profiles in Virginia and Maryland, vol 294. US Government Printing Office

Hack JT (1973) Stream-profile analysis and stream-gradient index. Journal of Research of the us Geological Survey 1(4): 421–429.

Haviv I, Enzel Y, Whipple KX, Zilberman E, Matmon A, Stone J, Fifield KL (2010) Evolution of vertical knickpoints (waterfalls) with resistant caprock: insights from numerical modeling. J Geophys Res Earth Surf 115(F3)

Hilley GE, Arrowsmith JR (2008) Geomorphic response to uplift along the Dragon's Back pressure ridge, Carrizo Plain. Calif Geol 36(5):367–370

Holbrook J, Scott RW, Oboh-Ikuenobe FE (2006) Base-level buffers and buttresses: a model for upstream versus downstream control on fluvial geometry and architecture within sequences. J Sediment Res 76(1):162–174

Holland WN, Pickup G (1976) Flume study of knickpoint development in stratified sediment. Bull Geol Soc Am 8:76–82

Lane EW (1955) Importance of fluvial morphology in hydraulic engineering. Proceedings American Society of Civil Engineers, 81, paper no 745

Leopold LB, Bull WB (1979) Base level, aggradation, and grade. Proc Am Philos Soc 123(3):168–202

Leopold LB, Langbein WB (1962) The concept of entropy in landscape evolution, vol 500. US Government Printing Office

Mackin JH (1948) Concept of the graded river. Geol Soc Am Bull 59(5):463–512

Pickup G (1977) Simulation modeling of river channel erosion. In: Gregory KJ (ed) River channel changes. Wiley, Chichester, UK, pp 47–60

Reed JC (1981) Disequilibrium profile of the Potomac River near Washington DC—a result of lowered base level or Quaternary tectonics along the fall line? Geology 9(10):445–450

Seidl MA, Dietrich WE, Kirchner JW (1994) Longitudinal profile development into bedrock: an analysis of Hawaiian channels. J Geol 102:457–474

Seidl MA, Dietrich WE, Schmidt KH, de Ploey J (1992) The problem of channel erosion into bedrock. Funct Geomorphol 23:101–124

Sklar LS, Dietrich WE (2008) Implications of the saltation–abrasion bedrock incision model for steady-state river longitudinal profile relief and concavity. Earth Surf Proc Land 33(7):1129–1151

Stock GM, Anderson RS, Finkel RC (2004) Pace of landscape evolution in the Sierra Nevada, California, revealed by cosmogenic dating of cave sediments. Geology 32(3):193–196

Stock GM, Anderson RS, Finkel RC (2005) Rates of erosion and topographic evolution of the Sierra Nevada, California, inferred from cosmogenic ^{26}Al and ^{10}Be concentrations. Earth Surf Proc Land 30(8):985–1006

Stolar DB, Willett SD, Montgomery DR (2007) Characterization of topographic steady state in Taiwan. Earth Planet Sci Lett 261(3–4):421–431

Wobus C, Whipple KX, Kirby E, Snyder N, Johnson J, Spyropolou K, Willett SD (2006) Tectonics from topography: procedures, promise and pitfalls. Special papers, Geological Society of America 398, p 55

Wolman MG (1987) Sediment movement and knickpoint behaviour in a small piedmont drainage basin. Geogr Ann 69A:5–14

Yanites BJ, Becker JK, Madritsch H, Schnellmann M, Ehlers TA (2017) Lithologic effects on landscape response to base-level changes: a modeling study in the context of the Eastern Jura Mountains, Switzerland. J Geophys Res Earth Surf 122(11):2196–2222

Rates

10

Abstract

Incision along a drainage network is controlled by the drainage area as well as by the channel slope and the erodibility. The deepest, steepest and narrowest incised valleys are formed when base-level fall is notably rapid. Increased precipitation accounts for increase in discharge, corresponding to an assumed increase in the erosional efficiency. Maximum incision rates occur at moderate sediment supply rates relative to the sediment transport capacity as a result of the tradeoff between the availability of abrasive tools and the partial alluviation of the bed. The use of cosmogenic isotopes analyses to determine surface exposure ages allows assessing the rates of retreat. Knickpoint migration retreat rates range <1 mm year^{-1} to a few tenths of mm year^{-1}, whereas a very fast rate may reach up to a few m/y. Fluctuation rates of base-levels controlled by tectonics are spread beyond the frequent <1–15 mm year^{-1} range, reaching 100 mm year^{-1} and beyond.

Keywords

Degradation index • Base-level drop rate • Incision rate • Retreat rate

10.1 Incision

Incision rate scales with stream power per unit area of the bed (Seidl and Dietrich 1992; Howard et al. 1994) and with the bed shear stress (Howard and Kerby 1983). The work done includes incision into bedrock and the transportation of the sediment load, i.e., the ability to physically detach and transport pieces of bedrock. River incision rate (E) can be defined as

$$E = KAmSn$$

where A is the upstream drainage area (a surrogate for discharge), S is the channel gradient, and power exponents m and n depend on basin hydrology, channel

D. Bowman, *Base-level Impact*,
https://doi.org/10.1007/978-3-031-24994-5_10

geometry and erosion processes. K is an erodibility constant which depends on the rock type over which the river channel flows as well as on the climatic setting. The rate of incision is thus enhanced by a relative low base-levels, high channel gradients, high regional precipitation and related discharge, relative erodible lithology and a narrow channel cross section. Incision rates peak at an intermediate level of excess shear stress due to a tradeoff between particle impact energy and impact frequency (Sklar and Dietrich 2004). Maximum incision rates occur at moderate sediment supply rates relative to the sediment transport capacity as a result of a tradeoff between the availability of abrasive tools and the partial alluviation of the bed.

Examples of incision rates

The Sila Massif in Italy (Fig. 10.1), an orogenic wedge seated on top of a narrow subduction zone, demonstrated erosion rates between 0.09 −/+0.01 and 0.13 −/+0.01 m/year.

The Indus River incises through the northwestern Himalaya at high rates of 2–12 mm year^{-1} (Burbank et al. 1996). In other Himalaya rivers, incision ranges similarly from a few millimeters per year in the Lesser Himalaya up to values of

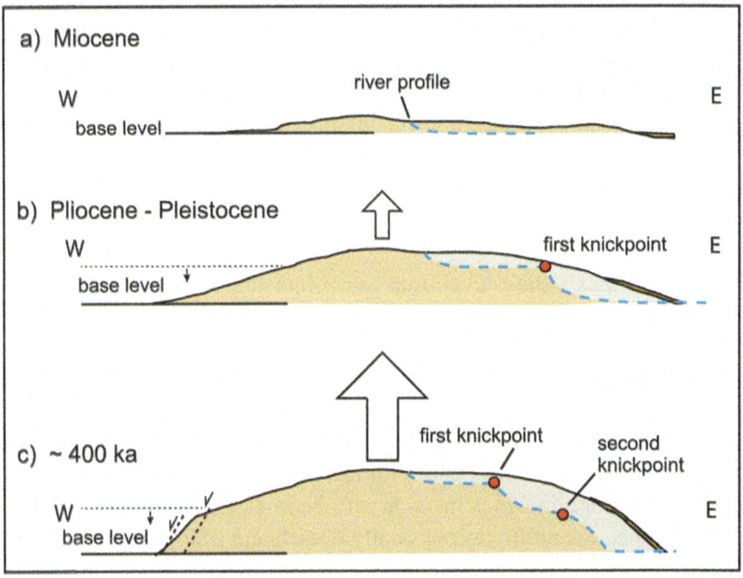

Fig. 10.1 Response of the Sila Massif, Italy, an orogenic wedge seated on top of a narrow subduction zone, to its uplift history through river valley evolution under falling base-level conditions, demonstrated schematically: **a** The high-standing low-relief surface on top of the massif has been interpreted as the relict of an old landscape. The strong incising rivers on the flanks of the massif (**b**, **c**) are characterized by two major knickpoints. It is suggested that the significant change in the rivers' profile shape, shown by steepening with time, is due to the abrupt increases in the uplift rate (From Olivetti et al. 2012)

10–15 mm year^{-1} (Whittaker et al. 2007). Long-term post-glacial incision rates between a minimum of 0.5 mm year^{-1} to a maximum of 15 mm year^{-1} have been inferred in sedimentary and crystalline rocks (Valla et al. 2010). Holocene extremely high incision rates of >20 mm year^{-1} were reported in gorges of the Swiss Alps (Korup and Schlunegger 2007). The lower reaches of the Yellow River are characterized by incision rates on the order of 1–2 mm year^{-1} (Harkins et al. 2007). Rockwell et al. (1984) reported incision rates of 0.8–2.2 mm year^{-1} through a highly resistant substrate at the Ventura River.

Tectonic base-level drops are base-level drops triggered by a *tectonic uplift.* Rock uplift rates are significant controlling factors of incision (Yoxall 1969; Schumm 1993; Heller et al. 2001). Rates of late Quaternary uplift at the northern termination of the San Andreas transform boundary in the USA, based upon dating the lowest, youngest raised marine terraces, spans an order of magnitude of 0.3–4 mm year^{-1}. Mendocino Triple Junction region in northern California shows, along the coast, Late-Pleistocene and Holocene very similar rock uplift rates that vary from 0.5 to 4 mm year^{-1} (Merritts and Bull 1989; Merritts and Vincent 1989). In the southwestern Tian-Shan, central Asia, the average rock uplift rates reach 1–3 mm year^{-1}. A rapid tectonic uplift rate, up to 15 mm year^{-1}, is shown at the Siwalik Hills of central Nepal. Base-level drop rates, reported by Attal et al. (2011), reveal that when controlled by tectonics, uplift may range 0.25–4.0 mm year^{-1} and even beyond 10 mm year^{-1}. This wide range of base-level fluctuation rates controlled by tectonics has been divided by Merritts et al. (1994) to *low* (<1 mm year^{-1}), *intermediate* (1–3 mm year^{-1}) and *high* (>3 mm year^{-1}).

Sediments control bedrock incision rate by providing tools for abrasion. The balance between river water discharge relative to the sediment concentration is a primary property in determining the stream erosional energy. Jansen et al. (2011) inferred that sediment availability, not water discharge, is the rate-limiting factor in knickpoint propagation for post-glacial knickpoints in western Scotland. Based on laboratory abrasion experiments (Sklar and Dietrich 2001), it had been suggested that sediment promotes erosion at low supply rates, but inhibits erosion at high rates by burying the underlying bedrock. *The highest erosion rate occurs at minimum bed cover and maximum bed impacts.* Jointing and dip orientation are important factors as well (Weissel and Seidl 1998). Fractures facilitate rapid knickpoint retreat rate by block plucking, i.e., through lifting of blocks defined by joints or fractures and bedding plans.

Relative incision or *degradation index* is the ratio of incision to the total base-level lowering suggested by Begin (1979). Under higher index values, incision migrates slowly upstream, while a smaller relative *degradation index* triggers a quicker migration (Fig. 10.2) *Channel narrowing or steepening* can lead to 2–tenfold increase in the incision rate (Hancock et al. 1998). Laboratory studies showed that the deepest, steepest and narrowest incised valleys are formed during the most *rapid fall of the base-level* (Begin et al. 1981; Schumm 1993; Blair and Mcpherson 1994). Valleys tend to widen when the rate of base-level fall decelerates as well as during base-level rise. Widening associated with erosion of the valley sidewalls acts to suppress valley-floor lowering.

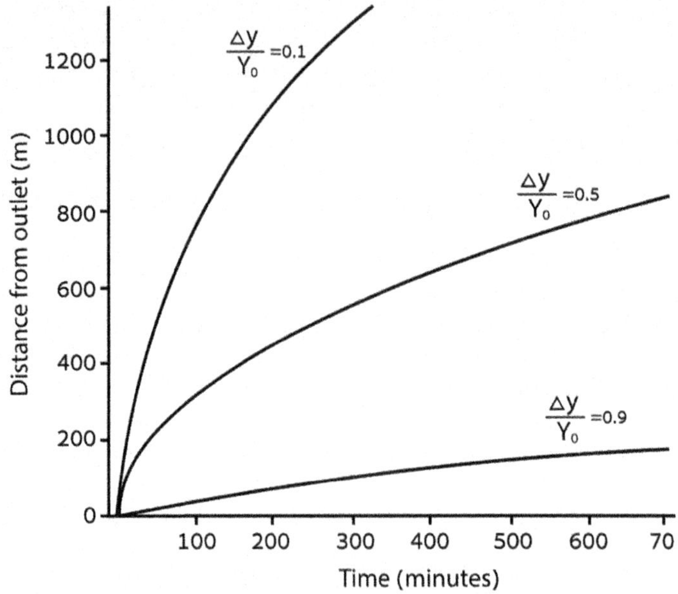

Fig. 10.2 Magnitude of degradation, defined as *degradation index* $\Delta Y/Y_0$ by comparing the elevation difference after degradation and before (ΔY) to the amount of base-level lowering (Y_0). Degradation index of different magnitudes propagates upstream at different rates with the smallest moving the fastest (From Begin 1979)

The main impact of erosion that follows base-level lowering is felt in the early stages mainly close to the outlet. The following responses to base-level lowering have been reported: Schumm (1979) showed that the rate of degradation at any station along the channel, after reaching its peak, slowly decreased as the channel became gradually adjusted to the new base-level. Experiments by Parker (1977) and by Hancock et al. (2002) showed that erosion following instantaneous base-level fall, started at high rates and then declined rapidly through time. Cantelli et al. (2004) reported high erosion rates during the initial stage of an experimental dam removal. Other experimental channel responses to base-level fall occurred very rapidly (within minutes), whereas the return to near-stabilization needed a time scale of several hours (Pelletier 2003).

10.2 Knickpoint Migration

The rate of knickpoint migration delimits the time necessary for the network to return to equilibrium after a base-level drop (Loget and Van Den Driessche 2009). The velocity of knickpoint retreat is inversely proportional to the substrate strength and a function of base-level fall rate. A main knickpoint may become combined with a smaller one, and their combined height becomes the factor controlling the retreat rate. The positive dependence of retreat rate on knickpoint height has been suggested by the experimental work of Holland and Pickup (1976) and by the numerical model of Haviv et al. (2010). However, field observations of Hayakawa and Matsukura (2003) suggest an inverse relation between retreat rate and knickpoint height. The retreat rate can be based as well on the ratio of the knickpoint-face height to flow depth (Fig. 7.2, Leopold et al. 1964). Retreat rates are often too slow for real-time measurement except in cases of alluvial sediments or where the underlying rock is especially weak.

The use of cosmogenic isotopes analyses to determine surface exposure ages allows to assess the rate of the retreat for hard bedrock knickpoints. The average rates over the entire retreat history can be estimated by dating the age of knickpoint initiation and measuring the distance of retreat. Knickpoint migration shows decreasing rates with time and distance upstream because of the smaller supplying drainage area and the resulting decline in discharge (Schumm et al. 1984). The upper declining basins of tributaries experience a smaller migration rate for a longer duration. Conclusions: the Knickpoint retreat is *drainage-area-dependent* (Hayakawa and Matsukura 2003; Crosby and Whipple 2006; Sklar and Dietrich 2004; Whipple and Tucker 2002; Turowski et al. 2007). The final result is the decreasing rate of upstream knickpoint propagation (Parker 1977). Experiments of Brush and Wolman (1960) and Leopold et al. (1964) showed similar exponential curves of decreasing rate of upstream knickpoint migration with time. However, upstream knickpoint migration may be also controlled by jointing and fracturing (Weissel and Seidl 1997, 1998) with the fluvial processes playing only a secondary role.

The final stages of knickpoint retreat may be very slow. Lague and Davy (2003) and Snyder et al. (2003) concluded that if the drainage area is below a critical area, at the order of 0.1–5 km^2 no fluvial processes will survive but mainly colluvial activity, debris flows and landsliding. Slowing of knickpoint retreat may also mark the gradual depletion of paraglacial sediment supply over the Holocene, leading to decrease in sediment availability and deficiency in tools for bedrock erosion (Sklar and Dietrich 2001).

Reported average knickpoint retreat rates for catchments with a drainage area between 6 and 65 km^2 ranged 0.2–2 mm year^{-1} (Whittaker and Boulton 2012). Modeled knickpoint migration rates activated by the upper Colorado River range up to 12 mm year^{-1}. Weissel and Seidl (1997) reported an escarpment retreat rate of 2 mm year^{-1}. Crosby and Whipple (2006) showed that knickpoint retreat velocities may be high and vary by a very wide range of magnitudes, i.e., 1–104 cm year^{-1}. Hayakawa and Matsukura (2003), Bishop et al. (2005), Lamb et al. (2007) and Valla et al. (2010) reported as well knickpoint retreat rates of several orders of magnitude, i.e., 0.5–2000 mm year^{-1}. Extreme knickpoint migration rates of 60–165 m year^{-1} were reported by Dente et al. (2017) in soft laminated lacustrine sediments composed of silt, halite and aragonite layers. The rate of recession of big waterfalls such as the Niagara and the Victoria Falls in Zambia reach up to 2 m year^{-1}. Additional knickpoint retreat rates in the range of 0.01–0.15 m year^{-1} were reported for different lithologies and climates over the world. A similar range, between 0.001 and 0.1 m year^{-1}, was suggested by Wohl (1998) and by Crosby and Whipple (2006).

Following deposition of lava flows, a new higher surface and drainage network develops, controlled by the initial, now relative lower, base-level (Zhang et al. 2011). Based on dating the new volcanic rocks, the initiation age of the new knickpoints was calculated and the recession rate reconstructed. Such knickpoint recession rates in the Yalu River-Wangtian'e volcanic region, China, were ~1–10 mm year^{-1} during the early stages of transient incision and became later ~1–6 mm year^{-1}.

Knickpoints have been found at the same distance upstream despite being very different in their nourishing drainage area, suggesting that inland propagation processes are not controlled only by drainage area/discharge ratio, but depends as well on complex additional controlling variables, such as the strength of the lithology and the ratio of the erosive force to the bedrock resistance. Climate may be responsible for a difference of one order of magnitude in the rate of knickpoint migration in bedrock rivers. Important may be as well the ratio of precipitation to infiltration (Whittaker and Boulton 2012). Bedrock erosion rates can vary in many orders of magnitude depending on the lithology. The effects of the rock structure should also not be ignored.

The rate of migration dictates the time needed for the drainage network to reach steady state (Gasparini et al. 2007; Neely et al. 2017). Very low retreat rates may form long-lived (>My) knickpoints. Field observations in accordance with theoretical considerations (Begin 1988) indicate that the distance of a knickpoint from the mouth of a tributary should be proportional to the square root of the time elapsed since the base-level was lowered. Uplift that started to trigger base-level fall several million years ago may have triggered erosion waves that are still propagating across the landscape. The propagation rates may not be steady and also fluctuate over timescales. A warmer and wetter post-glacial climate increases vegetation cover, provides greater soil cohesion and reduces sediment supply.

References

Attal M, Cowie PA, Whittaker AC, Hobley D, Tucker GE, Roberts GP (2011) Testing fluvial erosion models using the transient response of bedrock rivers to tectonic forcing in the Apennines, Italy. J Geophys Res: Earth Surf 116(F2)

Begin ZB (1979) Aspects of degradation of alluvial streams in response to base-level lowering. Unpublished Ph.D. dissertation, Colorado State University Fort Collins, Col, 239pp

Begin ZB (1988) Application of diffusion-erosion model to alluvial channels which degrade due to base-level lowering. Earth Surf Proc Land 13:487–500

Begin ZB, Meyer DF, Schumm SA (1981) Development of longitudinal profiles of alluvial channels in response to base-level lowering. Earth Surf Proc Land 6(1):49–68

Bishop P, Hoey TB, Jansen JD, Artza IL (2005) Knickpoint recession rate and catchment area: the case of uplifted rivers in Eastern Scotland. Earth Surf Processes Landforms: J Br Geomorphol Res Group 30(6):767–778

Blair TC, McPherson JG (1994) Alluvial fans and their natural distinction from rivers based on morphology, hydraulic processes, sedimentary processes, and facies assemblages. J Sediment Res 64(3a):450–489

Brush Jr LM, Gordon Wolman M (1960) Knickpoint behavior in noncohesive material: a laboratory study. Geol Soc Am Bull 71(1)

Burbank DW et al (1996) Bedrock incision, rock uplift and threshold hillslopes in the Northwestern Himalayas. Nature 379:505–510

Cantelli A, Paola C, Parker G (2004) Experiments on upstream-migrating erosional narrowing and widening of an incisional channel caused by dam removal. Water Resources Res 40(3)

Crosby BT, Whipple KX (2006) Knickpoint initiation and distribution within fluvial networks: 236 waterfalls in the Waipaoa River, North Island, New Zealand. Geomorphology 82(1–2):16–38

Dente E, Lensky NG, Morin E, Grodek T, Sheffer NA, Enzel Y (2017) Geomorphic response of a low-gradient channel to modern, progressive base-level lowering: Nahal HaArava, The Dead Sea. J Geophys Res Earth Surf 122(12):2468–2487

Gasparini NM, Whipple KX, Bras RL (2007) Predictions of steady state and transient landscape morphology using sediment-flux-dependent river incision models. J Geophys Res: Earth Surf 112(F3)

Hancock GS, Anderson RS, Whipple KX (1998) Beyond power: Bedrock river incision process and form. Geophys Monograph-Am Geophys Union 107:35–60

Hancock GS, Willgoose GR, Evans KG (2002) Testing of the SIBERIA landscape evolution model using the Tin Camp Creek, Northern Territory, Australia, field catchment. Earth Surf Processes Landforms: J Br Geomorphol Res Group 27(2):125–143

Harkins N, Kirby E, Heimsath A, Robinson R, Reiser U (2007) Transient fluvial incision in the headwaters of the Yellow River, northeastern Tibet, China. J Geophys Res: Earth Surf 112(F3)

Haviv I, Enzel Y, Whipple KX, Zilbermann E, Matmon A, Stone J, Fifield KL (2010) Evolution of vertical knickpoints (waterfalls) with resistant caprock: insights from numerical modeling. J Geophys Res 115(F3)

Hayakawa Y, Matsukura Y (2003) Recession rates of waterfalls in Boso Peninsula, Japan, and a predictive equation. Earth Surf Processes Landforms: J Br Geomorphol Res Group 28(6):675–684

Heller PL, Paola C, Hwang IG, John B, Steel R (2001) Geomorphology and sequence stratigraphy due to slow and rapid base-level changes in an experimental subsiding basin (XES 96–1). AAPG Bull 85(5):817–838

Holland WN, Pickup G (1976) Flume study of knickpoint development in stratified sediment. Geological Society of America Bulletin 87(1): 76–82

Howard AD, Dietrich WE, Seidl MA (1994) Modeling fluvial erosion on regional to continental scales. J Geophys Res: Solid Earth 99(B7):13971–13986

Howard AD, Kerby G (1983) Channel changes in badlands.Geol Soc Am Bull 94:739–75

Jansen JD, Fabel D, Bishop P, Xu S, Schnabel C, Codilean AT (2011) Does decreasing paraglacial sediment supply slow knickpoint retreat? Geology 39(6):543–546

Korup O, Schlunegger F (2007) Bedrock landsliding, river incision, and transience of geomorphic hillslope-channel coupling: evidence from inner gorges in the Swiss Alps. J Geophys Res: Earth Surf 112(F3)

Lague D, Davy P (2003) Constraints on the long-term colluvial erosion law by analyzing slope–area relationships at various tectonic uplift rates in the Siwaliks Hills (Nepal). J Geophys Res: Solid Earth 108(B2)

Lamb MP, Howard AD, Dietrich WE, Perron JT (2007) Formation of amphitheater-headed valleys by waterfall erosion after large-scale slumping on Hawaii. Geol Soc Am Bull 119(7–8):805–822

Leopold LB, Wolman MG, Miller JP (1964) Fluvial processes in geomorphology. W.H. Freeman and Co, San Francisco, 522p

Loget N, Van Den Driessche J (2009) Wave train model for knickpoint migration. Geomorphology 106(3–4):376–382

Merritts D, Bull WB (1989) Interpreting Quaternary uplift rates at the Mendocino triple junction, Northern California, from uplifted marine terraces. Geology 17(11):1020–1024

Merritts D, Vincent KR (1989) Geomorphic response of coastal streams to low, intermediate, and high rates of uplift, Medocino triple junction region, Northern California. Geol Soc Am Bull 101(11):1373–1388

Merritts DJ, Vincent KR, Wohl EE (1994) Long river profiles, tectonism and eustasy: a guide to interpreting fluvial terraces. J Geophys Res: Solid Earth 99(B7):14031–14050

Neely AB, Bookhagen B, Burbank DW (2017) An automated knickzone selection algorithm (KZ-Picker) to analyze transient landscapes: calibration and validation. J Geophys Res Earth Surf 122(6):1236–1261

Olivetti V, Cyr AJ, Molin P, Faccenna C, Granger DE (2012) Uplift history of the Sila Massif, southern Italy, deciphered from cosmogenic 10Be erosion rates and river longitudinal profile analysis. Tectonics 31(3)

Parker RS (1977) Experimental study of basin evolution and its hydrologic implication: unpublished Ph.D. dissertation, Colorado State University, Fort Collins, Colorado

Pelletier JD (2003) Drainage basin evolution in the rainfall erosion facility: dependence on initial conditions. Geomorphology 53(1–2):183–196

Rockwell TK, Keller EA, Clark MN, Johnson DL (1984) Chronology and rates of faulting of Ventura River terraces, california. Geol Soc Am Bull 95(12):1466–1474

Schumm SA (1979) Geomorphic thresholds: the concept and its applications. Trans Inst Br Geogr 485–515

Schumm SA (1993) River response to base-level change: implications for sequence stratigraphy. J Geol 101(2):279–294

Schumm SA, Harvey MD, Watson CC (1984) Incised channels: morphology, dynamics and control. Water Resources Publications

Seidl MA, Dietrich WE (1992) The problem of bedrock channel erosion. In: Schmidt K-H, DePloey J (eds) Functional geomorphology landform analysis and models. Catena Supplement, vol 2, pp 101–124

Sklar LS, Dietrich WE (2001) Sediment and rock strength controls on river incision into bedrock. Geology 29(12):1087–1090

Sklar LS, Dietrich WE (2004) A mechanistic model for river incision into bedrock by saltating bed load. Water Resources Res 40(6)

Snyder NP, Whipple KX, Tucker GE, Merritts DJ (2003) Channel response to tectonic forcing: field analysis of stream morphology and hydrology in the Mendocino triple junction region, Northern California. Geomorphology 53:97–127

Turowski JM, Lague D, Hovius N (2007) Cover effect in bedrock abrasion: a new derivation and its implications for the modeling of bedrock channel morphology. J Geophys Res: Earth Surf 112(F4)

Valla PG, van der Beek PA, Lague D (2010) Fluvial incision into bedrock: Insights from morphometric analysis and numerical modeling of gorges incising glacial hanging valleys (Western Alps, France). J Geophys Res 115

Weissel JK, Seidl MA (1997) Influence of rock strength properties on escarpment retreat across passive continental margins. Geology 25(7):631–634

Weissel JK, Seidl MA (1998) Inland propagation of erosional escarpments and river profile evolution across the Southeast Australian passive continental margin. Geophys Monograph-Am Geophys Union 107:189–206

Whipple KX, Tucker GE (2002) Implications of sediment-flux-dependent river incision models for landscape evolution. J Geophys Res: Solid Earth 107(B2):ETG-3

Whittaker AC , Boulton SJ (2012) Tectonic and climatic controls on knickpoint retreat rates and landscape response times. J Geophys Res: Earth Surf 117(F2)

Whittaker AC, Cowie PA, Attal M, Tucker GE, Roberts GP (2007) Bedrock channel adjustment to tectonic forcing: implications for predicting river incision rates. Geology 35(2):103–106

Wohl E (1998) Bed rock morphology and erosional processes. River over rocks. Fluvial processes in Bedrock channels. In: Thinkler, Wohl (eds). American Geophysical Union. Washington, DC

Yoxall WH (1969) The relationship between falling base-level and lateral erosion in experimental streams. Geol Soc Am Bull 80(7):1379–1384
Zhang HP, Zhang PZ, Fan OC 2011 Initiation and recession of the fluvial knickpoints: a case study from the Yalu River-Wangtian'e volcanic region, Nortwest Ern China. Sci China Earth Sci 54

The Messinian Event

11

Abstract

The *Messinian Salinity Crisis* of the Late Miocene was caused by reduced water inflow from the Atlantic Ocean to the Mediterranean Sea as a result of a tectonic uplift of the Gibraltar arc and global sea level changes. The consequent rapid and huge Mediterranean base-level drop of about 1500 m triggered an extensive subaerael erosion along the exposed Mediterranean margins, thereby demonstrating a dramatic base-level impact. The Messinian base-level fall initiated deep incision and rapid headward rejuvenation along gigantic canyons with several hundred meters depth between interfluves and the channel beds which are visible in numerous seismic reflection profiles in a variety of physiographic settings around the Mediterranean. The Rhone, Po and the Nile serve as examples for such deep incised paleovallies that propagated far inland, up to several hundreds km. Additional markers of the huge phase of the subaerial erosion following the Messinian base-level drawdown include erosional discordances, submarine scarps, submarine canyons and platforms, mesa-like relief and valley terraces.

Keywords

Messinian erosional surface • Submarine paleovalley • Proto Nile canyon

11.1 The Base-Level Drop Event

For improving understanding base-level impacts, we can use the intense researched example of an abrupt base-level fall event that occurred in the Mediterranean, known as the *Messinian Salinity Crisis* of Late Miocene. This is considered the most dramatic base-level fluctuation ever recorded in geological times (Ryan et al. 1973). It was caused by reduced water inflow from the Atlantic Ocean into the Mediterranean Sea, resulting in a sharp decrease in the Mediterranean sea level due to evaporation. The reduced connectivity between the Atlantic Ocean and the Mediterranean is thought to have resulted from tectonic uplift of the Gibraltar arc

D. Bowman, *Base-level Impact*,
https://doi.org/10.1007/978-3-031-24994-5_11

seaway and from global sea level changes which both controlled the inflow of water and triggered the rapid and dramatic environmental change (Hsü et al. 1973; Loget and Van Den Driessche 2009).

The dramatic sea level drop of the *Messinian Salinity Crisis* started in Late Miocene with the progressive closure of the connection between the Atlantic Ocean and the Mediterranean Sea (Gautier et al. 1994; Krijgsman et al. 1999) at about 5.96 Ma ago. This event that closed the Atlantic gateways of the Betic-Rifian corridors ended at 5.33 Ma, spreading the Messinian Salinity event along a period of ~0.63 Ma (Hilgen et al. 2007). Isolation of the Mediterranean Sea from the Atlantic Ocean caused an extreme sea level fall of about 1500–2000 m and gave rise to precipitation of evaporites over the desiccated Mediterranean basins (Hsü et al. 1973; Lofi et al. 2005). The Messinian sea level drop is presented here as a mega base-level event with enormous morphological impacts. Discussion of the open problems related to the Messinian crisis is beyond the scope of this work.

The rapid, km-scale, Messinian base-level lowering event, caused extensive sub-aerael erosion along the exposed basin margins and involved deposition of thick evaporites in the deep basins. Simultaneously, a wave of river incision propagated along the Mediterranean margins inland with a rate that varied as a function of the drainage areas. The Messinian event has been traditionally envisaged as a rapid event (Hsü et al. 1973) that ended abruptly with the rapid reflooding of the desiccated Mediterranean basin during the early marine Pliocene transgression (Loget et al. 2005) by a fast sea level rise, up to a few tenth of meters above the present sea level. Reflooding through the Strait of Gibraltar was triggered primarily by tectonic subsidence at the Gibraltar sill which allowed the return of open marine conditions and full circulation within the Mediterranean Sea. The Pliocene reflooding and infilling buried the exposed erosional Messinian morphology and thus allowed its preservation.

The *Messinian Erosional Surface* or the *Messinian Discontinuity* (Fig. 11.1) is the discordance of the Miocene with the following Pliocene deposits (Lofi and Berné 2008; Ryan and Cita 1978). This surface is sandwiched between Miocene deposits and the first Pliocene deposit and can be clearly traced on seismic profiles to depths of 2.5–5.0 s (two-way travel time).

11.2 Morphological Impacts

The presence of gigantic Messinian canyons along the continental margins of Mediterranean basins filled by Pliocene sediments confirm the deep drawdown of the Messinian base-level. Erosional Messinian discordances are visible on numerous seismic reflection profiles in a variety of physiographic settings (Fig. 11.2). The Messinian relief is rugged as badland topography, often similar to channel and gully morphology (Stampfli and Höcker 1989; Frey Martinez et al. 2004). Incision rates were considerable, up to 10 mm year^{-1}, similar to fluvial incision rates of 2–12 mm year^{-1} in the tectonically active mountain belts of the Himalayas (Burbank et al. 1996).

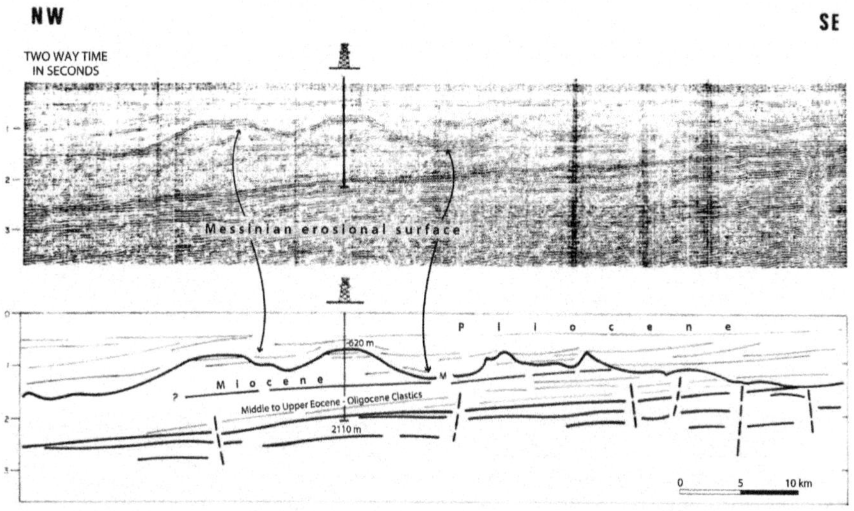

Fig. 11.1 Seismic section and its interpretation documenting the dramatic Messinian base-level fall by showing the *Messinian subaerial erosional surface* as a major unconformity. The Messinian erosional surface is bevelling the underlying strata and can be detected as discordance on numerous seismic reflection profiles (modified after Barber 1981)

The markers that confirm the aerial erosional activity related to the Messinian event are morphological features determined mainly from submersible dives (Savoye and Piper 1991), conventional and industrial seismic reflection profiles (Lofi et al. 2005) and from integration of high-resolution marine seismic profiles and analysis of sea-bottom cores (Tassy et al. 2014). The markers include: submarine scarps, submarine platforms, mesa-like relief, filled submarine canyons, valley terraces, irregular high-relief surfaces, extremely rough and steeply gullied morphologies analogous to badlands, paleochannels which are part of continuous ancient drainage nets, large canyons, large-scale "cross-bedding" and abundance of "cut and fill" structures (Ryan and Cita 1978). The submarine thalwegs of the canyons are sometimes occupied by huge, longitudinal chaotic piles of blocks, partly buried in sediments that occupy the whole width of the thalweg. Thick, sandy and conglomeratic layers commonly display slump structures typical of high-energy deposits. Accumulation of these detrital successions in distal deeper parts of submerged drainage networks mark the subaerial erosion that followed the base-level drawdown (Bache et al. 2009).

The dramatic Messinian base-level fall led to the deep incision and uphead erosion along huge canyons around the Mediterranean (Loget et al. 2005) detected on the Catalan and Spain margins (Maillard et al. 2006), on the southern margin of France, in the southwestern and southeastern Tyrrhenian Sea, on the margin of the Levantine coast and in the Aegean Sea. The Messinian event affected as well the incision of the Po basin that propagated very far inland, up to several hundreds km, extending into the Southern Alps and causing a northward shift of the watershed.

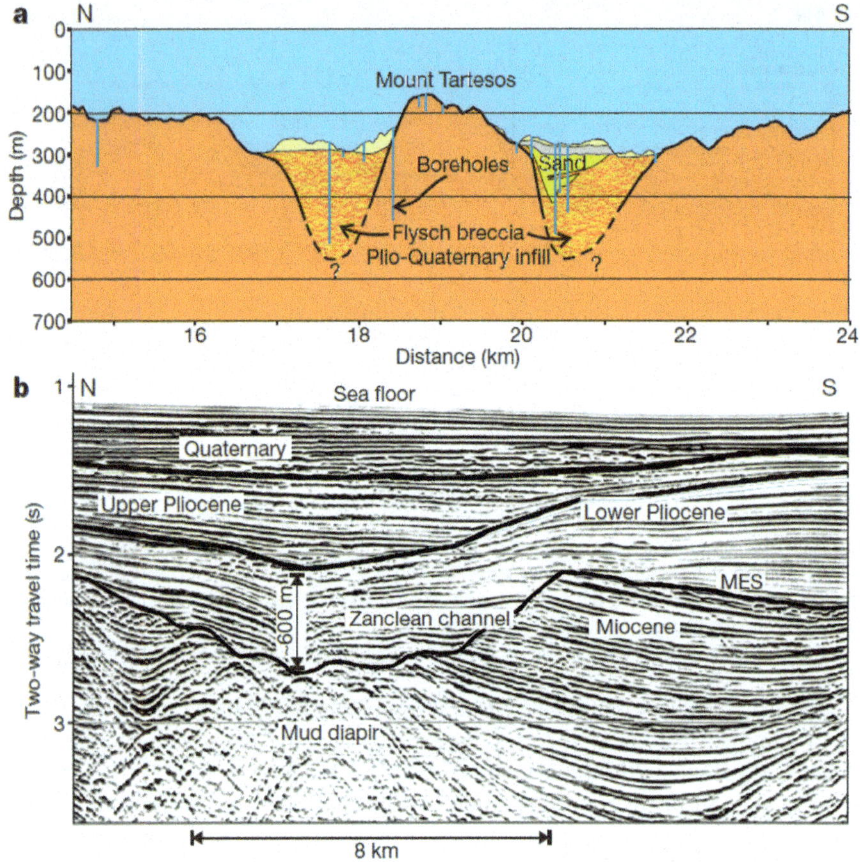

Fig. 11.2 Messinian erosive morphology based on stratigraphy provided by seismic profiles and borehole explorations: **a** erosive troughs across the Gibraltar strait with flysch fill. Vertical exaggeration 5:1. **b** An incised channel as part of the Messinian Erosional Surface (MES). Vertical exaggeration 4:1 (from Garcia-Castellanos et al. 2009). The short Messinian time interval must have been long enough to incise deep channels buried later by sediments and noticed in seismic reflection profiles around the Mediterranean (Loget and Van Den Driessche 2009; Rizzini et al. 1978)

The Alpine rivers carved up to 1 km deep V-shaped valleys across the modern Po Basin and incised far into the Alp by regressive erosion creating deep valleys that confine today the glacial Alpine Lakes of Northern Italy.

The Nile River has cut a deep channel upstream to Aswan (Chumakov 1967) where Messinian vertical incision of nearly 80 m has been identified 1000 km south from the present coast. The Proto Nile canyon is approximately 12 km wide, marking the attempt to regain a *Messinian equilibrium*. The geometric reconstructed profile of the Messinian Rhone paleovalley in France, based on seismic and drilling data with field observations, when compared to the ante-Messinian profile (Figs. 11.3 and 11.4) shows dramatic incision (Gargani 2004) extending 300 km upstream from the sea, filled with Pliocene marine sediments. Along the Catalan-continental margins, numerous submarine valleys have been eroded forming a convergent network of tributaries that flows down into the deep Valencia paleo-channel (Garcia et al. 2011). The erosional systems include up to fifth-order dendritic drainage networks (Gorini et al. 2005) with several hundred meters depth between interfluves and the channel bed.

The northward shift of the watershed in the Alps captured parts of the former northward draining networks, resulting in extension of the drainage from the Alps toward the south which continues today. The recent drainage of part of the Southern French Pyrenean became controlled as well by the low Messinian base-level. It had been suggested that glaciers that overdeepened alpine valleys were preconditioned by the Messinian incision and that major knickpoints that can be observed in surface streams originated from relic Messinian channels (Robl et al. 2017). Along the Tyrrhenian coast of the Gulf of Genoa, as well as along the entire western Mediterranean, coastal-erosional-paleo-valleys have been recorded.

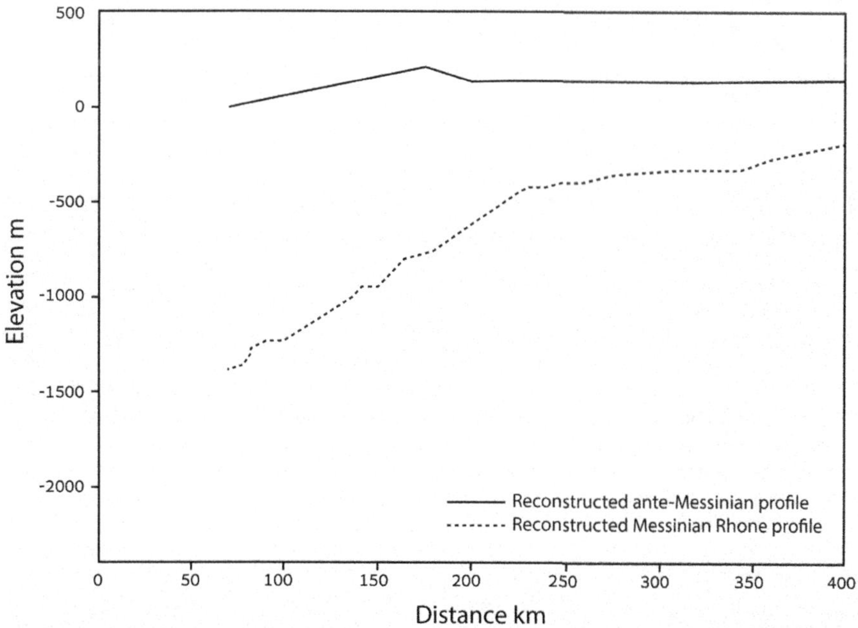

Fig. 11.3 Reconstructed ante-Messinian Rhone river profile compared to the reconstructed Messinian profile taking into account post Messinian tectonics and isostatic deformations due to valley incision and sea unloading. Note max. base-level difference of about 1500 m (after Gargani 2004)

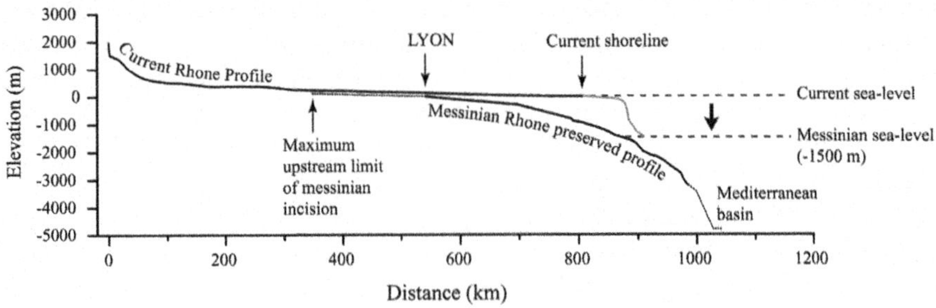

Fig. 11.4 Messinian Rhone river profile controlled by the 1500 m low base-level. The Messinian profile demonstrates a deeply entrenched and infilled Miocene valley, underlying the recent Rhone valley. The convex shape indicates disequilibrium (from Loget et al. 2006)

References

Bache F, Olivet JL, Gorini C, Rabineau M, Baztan J, Aslanian D, Suc JP (2009) Messinian erosional and salinity crises: view from the Provence Basin (Gulf of Lions, Western Mediterranean). Earth Planet Sci Lett 286(1–2):139–157

Barber PM (1981) Messinian subaerial erosion of the proto-Nile Delta. Mar Geol 44(3–4):253–272

Burbank DW, Leland J, Fielding E, Anderson RS, Brozovic N, Reid MR, Duncan C (1996) Bedrock incision, rock uplift and threshold hillslopes in the northwestern Himalayas. Nature 379(6565):505–510

Chumakov IS (1967) Pliocene and Pleistocene deposits of the Nile Valley in Nubia and Upper Egypt. Dokl Akad Nauk SSSR 170:5 (in Russian)

Garcia-Castellanos D, Estrada F, Jiménez-Munt I, Gorini C, Fernández M, Vergés J, De Vicente R (2009) Catastrophic flood of the Mediterranean after the Messinian salinity crisis. Nature 462(7274):778–781

Garcia M, Maillard A, Aslanian D, Rabineau M, Alonso B, Gorini C, Estrada F (2011) The Catalan margin during the Messinian salinity crisis: physiography, morphology and sedimentary record. Mar Geol 284(1–4):158–174

Gargani J (2004) Modelling of the erosion in the Rhone valley during the Messinian crisis (France). Quatern Int 121(1):13–22

Gautier F, Clauzon G, Suc JP, Cravatte J, Violanti D (1994) Age and duration of the Messinian salinity crisis. CR Acad Sci Paris (IIA) 318:1103–1109

Gorini C, Lofi J, Duvail C, Dos Reis T, Guennoc P, Le Strat P, Mauffret A (2005) The Late Messinian salinity crisis and Late Miocene tectonism: interaction and consequences on the physiography and post-rift evolution of the Gulf of Lions margin. Mar Pet Geol 22:695–712

Hilgen F, Kuiper KF, Krijgsman W, Snel E, Van Der Laan E (2007) Astronomical tuning as the basis for high resolution chronostratigraphy: the intricate history of the Messinian salinity crisis. Stratigraphy 4:231–238

Hsü KJ (1973) The desiccated deep-basin model for the Messinian events. In: Drooger CW (ed) Messinian events in the Mediterranean. North-Holland Publ. Co., Amsterdam, pp 60–67

Hsü KJ, Ryan WB, Cita MB (1973) Late Miocene desiccation of the Mediterranean. Nature 242(5395):240–244

Krijgsman W, Hilgen FJ, Raffi I, Sierro FJ, Wilson DS (1999) Chronology, causes and progression of the Messinian salinity crisis. Nature 400:652–655

Lofi J, Déverchère J, Gaullier V, Gillet H, Gorini C, Guennoc P, Obone Zue Obame E (2008) The Messinian Salinity Crisis in the offshore domain: an overview of our knowledge through seismic profile interpretation and multi-site approach. In CIESM Workshop Monographs Vol. 33: 83–90

Lofi J, Gorini C, Berné S, Clauzon G, Dos Reis AT, Ryan WB, Steckler MS (2005) Erosional processes and paleo-environmental changes in the Western Gulf of Lions (SW France) during the Messinian salinity crisis. Mar Geol 217(1–2):1–30

Loget N, Davy P, Driessche JVD (2006) Mesoscale fluvial erosion parameters deduced from modeling the Mediterranean sea level drop during the Messinian (Late Miocene). J Geophys Res Earth Surf 111(F3)

Loget N, Driessche JVD (2009) Wave train model for knickpoint migration. Geomorphology 106(3–4):376–382

Loget N, Driessche JVD, Davy P (2005) How did the Messinian salinity crisis end? Terra Nova 17(5):414–419

Maillard A, Gorini C, Mauffret A, Sage F, Lofi J, Gaullier V (2006) Offshore evidence of polyphase erosion in the Valencia Basin (Northwestern Mediterranean): scenario for the Messinian salinity crisis. Sed Geol 188:69–91

Martinez JF, Cartwright JA, Burgess PM, Bravo JV (2004) 3D seismic interpretation of the Messinian Unconformity in the Valencia Basin, Spain. Geological Society, London, Memoirs, 29(1):91–100

Rizzini A, Vezzani F, Cococcetta V, Milad G (1978) Stratigraphy and sedimentation of a Neogene—quaternary section in the Nile Delta area (ARE). Mar Geol 27(3–4):327–348

Robl J, Heberer B, Prasicek G, Neubauer F, Hergarten S (2017) The topography of a continental indenter: the interplay between crustal deformation, erosion, and base-level changes in the eastern Southern Alps. J Geophys Res Earth Surf 122(1):310–334

Ryan WBF, Cita MB (1978) The nature and distribution of Messinian erosional surface—indication of a several kilometer-deep Mediterranean in the Miocene. Mar Geol 27:193–230

Ryan WBF, Hsu KJ et al (1973) Initial reports of the Deep Sea Drilling Project, vol 13. US Government Printing Office, Washington DC, 1447

Savoye B, Piper DJW (1991) The Messinian event on the margin of the Mediterranean Sea in the Nice area, southern France. Mar Geol 97:279–304

Stampfli GM, Höcker CFW (1989) Messinian palaeorelief from a 3-D seismic survey in the Tarraco concession area (Spanish Mediterranean Sea). Geol Mijnb 68:201–210

Tassy A, Fournier F, Munch P, Borgomano J, Thinon I, Fabri MC, Cornee JJ (2014) Discovery of Messinian canyons and new seismic stratigraphic model, offshore Provence (SE France): implications for the hydrographic network reconstruction. Mar Pet Geol 57:25–50

Morphological Products

12

Abstract

Knickpoints mark the upper boundary of a channel reach that is in control of a base-level. Beyond the knickpoint upstream, the impact of the falling base-level has not yet been transmitted. This makes a typical *two-storey morphology* based on differences in gradients, incision rates, cross-sectional morphology, in the abundance of terraces as well as in slope dynamics. A river terrace designates a low base-level that has initiated an upstream propagating wave of incision. Paleo-longitudinal channel profiles can be reconstructed based on treads of terraces. Downstream converging, diverging or parallel profiles, reconstructed by terraces, reflect different dynamics of their controlling base-level and may offer an estimate of the trend and magnitude of the base-level dynamics. Channels are bounded from above by the highest terrace and from below by the deepest incised channel, portraying altogether a three dimensional envelope of aggradation and degradation history including repeated cut-and-fill cycles. Coupling between incising channels and hillslopes plays an important role in the relief evolution. Quantifying landslides shows that the frequency of slope failures per unit area is largest in zones that experienced the control of a falling base-level. Channels undercutting toes of hillslopes and initiating slope oversteepening provide the conditions for slope adjustment. Hillslope sediment flux and landsliding may be solely driven by the rates of base-level fall and the fluvial incision.

Keywords

Transient state • Inner gorge • Two-storey morphology • Hillslope–channel coupling • Divide migration • Strath terrace • Piracy • Relief evolution

D. Bowman, *Base-level Impact*,
https://doi.org/10.1007/978-3-031-24994-5_12

12.1 Two-Storey Morphology

Before a complete adjustment of the drainage network to the base-level is reached, the channel network is in a *transient state*, responding and adjusting to the base-level fall through a passing wave of incision (Zhang et al. 2017). A knickpoint marks the upper boundary of the channel reach that is adjusting to the falling base-level, the impact of which is not transmitted upstream beyond such knick-point (Fig. 12.1). Transient conditions can persist within the drainage network up to several tens of 10^6 y (Clark et al. 2005) during which two diagnostic morphological compartments evolve, i.e., a *two-storey morphology* that represents the pre- and post-passage of the erosional wave.

Floor A—is adjusting toward the lowered base-level with steep gradients, often staying for very long periods incompletely adjusted. The channel starts to adjust closest to the outlet which is most affected by the base-level fall (Bevis 2015) and where the erosion rates are the highest, a few times faster than in **floor B** above, which is still ignorant of the downfloor effective incision and eroding at very low rates (Carretier and Lucazeau 2005; Weissel and Seidl 1997). Morphologically floor A is a deep, narrow, "v"-shaped gorge with slopes stripped of soil, demonstrating a *valley-in-valley, inner gorge* morphology (Fig. 6.3).

Inner gorges are flanked by rectilinear and occasionally subvertical bedrock cliffs. Narrowing is an intrinsic way by which the system maximizes its erosional response. Incision is communicated to the hillslopes, resulting by landsliding with high bluff frequency. Hillslopes above the inner gorge walls and the inflection point may stay largely decoupled from the incision process in the inner gorge (Korup and Schlunegger 2007). This limited geomorphic coupling of the upper hillslopes provides evidence of the transient nature of inner gorges.

Density of large clasts is high along inner gorges compared to the upstream surface (floor B). Channels at floor A are typically mantled with large boulders which inhibit further channel incision before being broken down during a large flood, or removed through undermining. The uppermost transient boundary of the incised A segment is a knickpoint or a convex knickzone (Crosby 2006; Crosby and Whipple 2006; Harkins et al. 2007; Reinhardt et al. 2007).

Floor B—includes the channel reach upstream, above the knickpoint. It was in equilibrium with former base-level conditions. This upper landscape is distinguishable from the downstream more equilibrated A area by the hillslope morphology, the channel gradients, the dominant erosional processes and the erosion rates. On the B floor, above the knickpoint, the area shows gentle hillslopes of a slowly eroding, quasi-equilibrium landscapes (Whittaker et al. 2007). The A floor gradients are almost 2–2.5 times steeper (Morell et al. 2012). The topography of floor B has lower local relief with relatively wider floodplains. Transects through valleys in floor B are bowl-shaped with soil-mantled valley walls which are notably different from the V-shaped cross sections of floor A (Fig. 12.2).

On floor B, incision rates are in equilibrium with former base-level conditions (Whipple et al. 2013) and are not responding to the recent, active base-level fall (Clark et al. 2006). Floor B is typified by low-relief, smooth slope gradients, slow

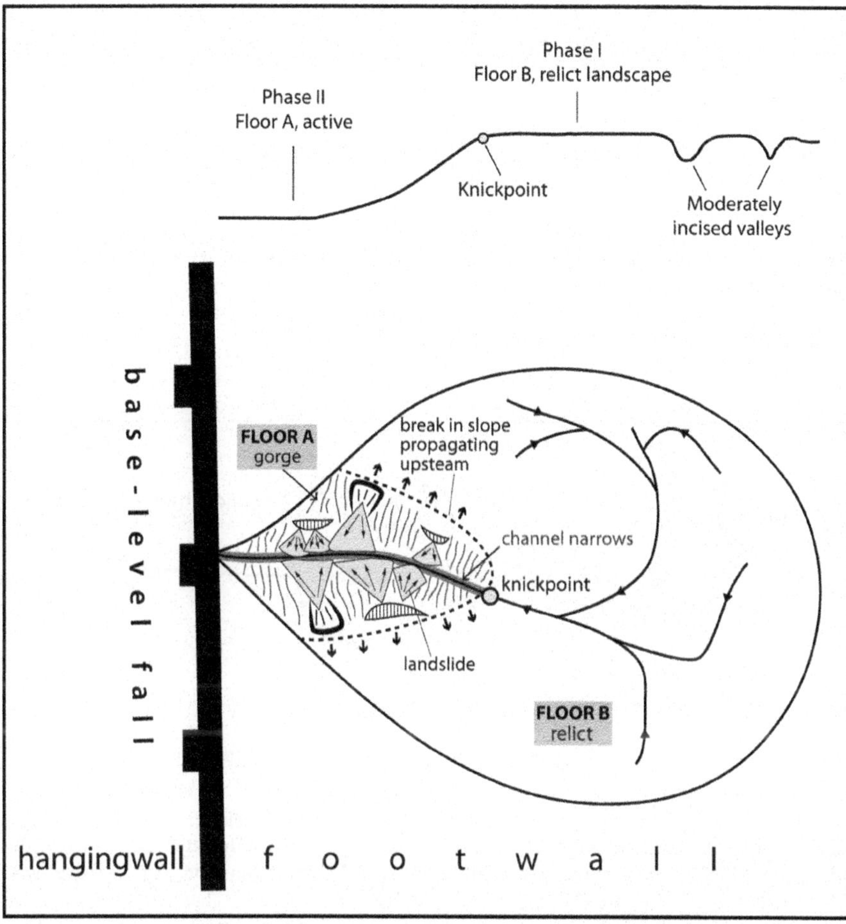

Fig. 12.1 Sketch illustrating a footwall uplift (base-level fall) that initiated incision and formed a *two-storey morphology*: the recent, active lower Floor **A** demonstrates a narrow transient gorge morphology propagating upstream, typified by steep, active slopes, cliffs, high incision rates, landslides, scree cones, coarse sediment supply and terraces. The upper relict Floor **B** composes a paleo-land, ignorant of the high incision rates downfloor. The upper floor includes the drainage network above the knickpoint, and its relict relief is typified by gentle graded slopes, soil cover, finer sediments and lower erosion rates (Modified from Attal et al. 2011)

erosion rates and broader valley bottoms. The inequality between the moderate channel incision rates on floor B and the recent base-level fall is great. The conclusion that floor B is unaware of the recent base-level fall suggests that incision has not yet progressed entirely through the drainage network (Clark et al. 2006). The paleo-higher base-level, toward which floor B surface had responded, can be reconstructed, based on terraces (Fig. 12.3). Such reconstruction will provide as well the amount of incision below the upper relict B surface.

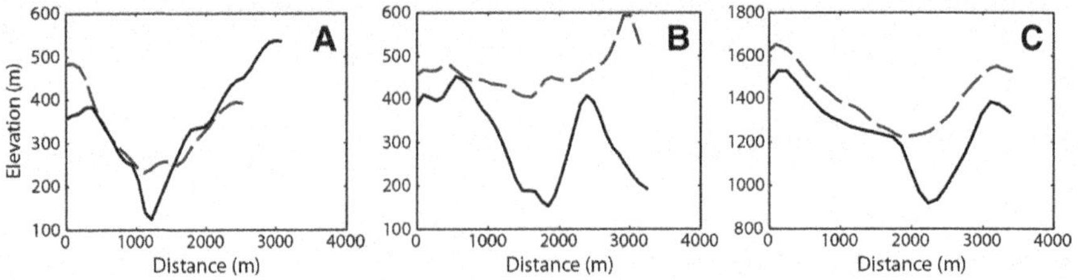

Fig. 12.2 Cross-valley profiles through examples of a *two-storey morphology* in eastern Papua New Guinea. The upper dashed line shows Floor B—a relict, smooth morphology with no indication for base-level impact. The lower solid line represents Floor A—the inner gorge following intensive incision (From Miller et al. 2012)

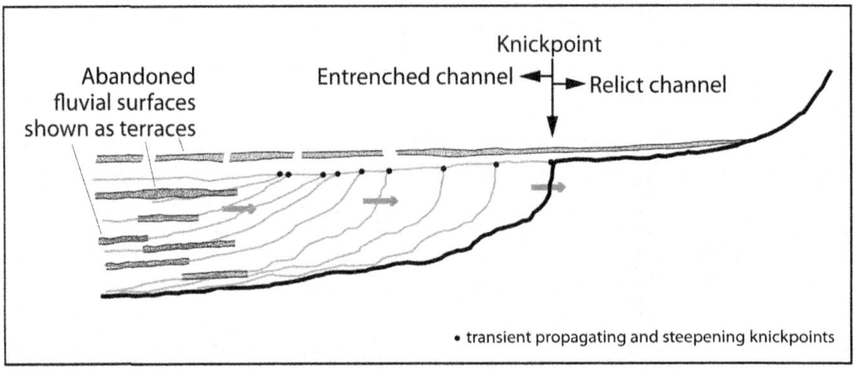

Fig. 12.3 Schematic illustration of a typical transient propagating knickpoint and its reconstructed paleo-profiles. The knickpoints define the mobile boundary between the adjusted and entrenched valley segment controlled by the recent falling base-level (area A) and the relict, perched channel above the knickpoint (area B). Fluvial terraces are relicts of paleo-channels. One terrace can be reconstructed off the lip of the active knickpoint (Modified after Crosby and Whipple 2006). To estimate the magnitude of a base-level fall we should reconstruct the highest relict profile and compare it to the modern active channel. The difference in elevation provides constraints on the maximum base-level fall. Channels downstream of knickpoints are similar in their erosional rate in many tectonic active orogens worldwide, including the Himalaya (Wobus et al. 2003), eastern Tibet (Ouimet et al. 2009) and the San Gabriel Mountains (DiBiase et al. 2015). All are in an eroding range of 1–3 mm/year (Morell et al. 2012)

The striking contrast between the slowly eroding upper terrain (level B) and the rapidly wearing away lower floor (level A) is demonstrated by the Yellow River. Its lower reaches are characterized by incision rates in the order of 1–2 mm/year, i.e., a few hundreds of meters per Ma^{-1}, while tributary reaches upstream of knickpoints incise at much slower rates, i.e., between 0.05 and 0.1 mm/y, ~50–100 m Ma^{-1} (Harkins et al. 2007). The lower part of the Yellow River exhibits steeper channels, narrow gorges, steep coupled hillslopes and high flow velocity as expected.

An additional example of the two distinct geomorphic zones has been reported in a 27 km^2 catchment in southern Spain (Reinhardt et al. 2007). The steep catchment below the knickpoint showed rapid channel incision of 5 ± 1 mm/y over 12 ka, whereas in the upland relatively low erosion rates of only 0.05 ± 0.02 mm/y have been reported. Prince and Spotila (2013) analyzed knickpoints and paleo-landscapes of the New River basin, the southern Appalachians. They concluded that where base-level drop is rapid and significant, increased stream power and bedrock incision induces a *transient state* of incision (area A), i.e., a signal that migrates upward the drainage network, bounded by steep-walled gorges, knickpoints and steepened bedrock slopes that are the headward-mobile response to the base-level drop. The capacity of hillslopes to provide sediment to the energized drainage network was great in this transient erosive areas. The A area was too rugged and lacking in soil cover to permit agriculture, which is limited to the gentler topography of the elevated relict surface (area B).

Brocard et al. (2015) suggested that the increase in erosion rates that accompany the passage of erosional waves makes knickpoints a major boundary between an upstream forest that retrieves most of its nutrients from the atmosphere and the downfloor that retrieves a significant fraction of nutrients from the bedrock, demonstrating the importance of the base-level on the structure of an old-growth tropical forest. The passage of a headward retreating knickpoint is thus suggested as a first-order effect on the delivery of nutrients to the overlying rainforest.

12.2 River Terraces

Terraces are residuals of a former channel bed, that is to say, the result of incision into the river bed accompanied by lateral migration. Terraces are to be found where the stream power was in excess of that needed to move sediments. It has become almost axiomatic to look on terraces as recording incisional conditions following faulting and folding, i.e., an indication of a falling base-level (Merritts and Vincent 1989; Merritts et al. 1994; Mather et al. 2017). Formation of terraces as a time-transgressive process can be triggered by an increase in rock uplift rate which can be conceived as change in the rate of base-level fall, or by an increase in the erosional efficiency following a climatic event. Vertical incision rate can be determined from heights and ages of fluvial terraces. The terraces are thus most common in lower reaches where incision is greatest and extend downstream from knickpoint crests (Fig. 12.4). The absence of terraces often reflects lack of preservation.

The number of base-level drops can be partly reconstructed by the number of the terraces, although not all terraces may be preserved. Along steep gorges, the terrace record is often lost. In a flume experiment, several terrace levels have been formed during incision as response to only one drop in the base-level (Gardner 1983). Perhaps a few unrecorded pulses could explain this phenomena.

Paleo-longitudinal profiles can be reconstructed and mapped based on treads of terraces, although often scarce. *Strath terraces* are surfaces cut into bedrock,

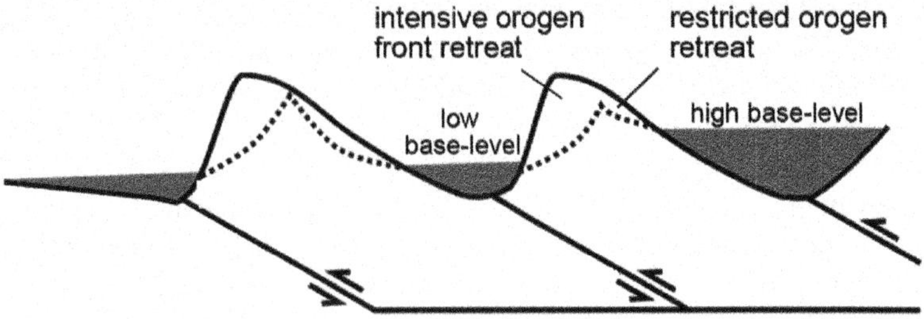

Fig. 12.4 Asymmetric uplift of a mountain range may create significant base-level altitude differences, or difference in the distance of the base-levels from the divide. The morphology of a range will be controlled by competition between the geometry of the underlying faults and the base-levels that flank the emerging range. The divides will retreat under the focused rapid erosion on the steeper flank. The base-level altitudinal contrasts modify the asymmetric tectonic morphology too (from Densmore et al. 2005, Modified after Barnes et al. 2011)

products of paleo-lateral river cut. Strath terraces can be capped with a few m of channel conglomerate deposits and soil that allow dating. The alluviation may develop as a result of imbalances between local transport capacity and upstream sediment supply. The Alluvium capping bedrock marks part of the cut and fill cycle which forms the terraces.

The ratio of lateral erosion to vertical incision varies due to different uplift rates (base-level falls) and magnitudes of the upstream drainage area. The typical river terraces are expected to form following lateral incision during a relatively slow and continuous decelerating base-level fall. Multiple terrace-pairs are expected during a constant rate of base-level fall (Muto and Steel 2001). As the degree of adjustment to a base-level depends upon the size and power of the river, adjacent rivers of different sizes with the same uplift rate of the area are not expected to have similar numbers of terraces. Terraces in the same drainage basin must as well not grade to the contemporaneous base-level but may reflect former, different base-level altitudes.

12.3 Hillslope Control

Although hillslope processes and failures may be caused by intense precipitation or seismic shaking, unrelated to base-level, interaction between channel incision and hillslopes plays a significant role in relief evolution. Changes in hillslope morphology and soil-mantle properties can be driven by channel incision at the base of slopes triggered by a falling base-level (Mudd and Furbish 2007). The upstream propagation of knickpoints improves the linkage between fluvial and hillslope processes.

Based on small-scale erosion experiments in controlled drainage basins, Bigi et al. (2006) showed that hillslopes can be cleared and material transported away from steep slope toes by sufficient stream power. Coarse material pulled down and delivered into the channel can become bedload that increases the incision rate by plucking. Supply of more hillslope sediments, when the fluvial water discharge remains constant, may exceed the transport capacity of the channel and lead to deposition and shielding of the bed, inhibiting further incision (Sklar and Dietrich 1998).

As channels steepen and narrow the adjacent hillslope tend to steepen. Numerous hillslope soil properties as soil thickness and soil production change with changes in the steepness of the hillslope. As steepening of slopes leads to gradual removal and thinning of the soil veneer, soil records provide a window into the entrenchment history. When bedrock incision keeps pace with rock uplift (equilibrium), the valley sides will stay at a critical slope angle for rock failure. Thus, steep hillslopes often indicate slope adjustment to channel incision triggered by base-level lowering.

Fluvial and landsliding processes will be tightly coupled when hillslopes keep pace with channel incision (Densmore et al. 1998; Gallen et al. 2011). A continued base-level fall may thus mean that landslide-prone hillslopes are continuously created and deliver sediments. However, mass movement processes as shallow landsliding or solifluction often do not occur on the upper portions of hillslopes which can largely remain decoupled from the channels, i.e., left without any fluvial forcing (Dapples et al. 2003).

When slopes are maintained near equilibrium, only a short response time is needed for them to react to the channel incision at their base. Hillslopes of larger river basins may be able to respond essentially instantaneously to base-level fall by slope failures and mass wasting events which will delay further channel entrenchment and thereby respond negatively to further knickpoint propagation (Weissel and Seidl 1998). Conversely, it may take for hillslopes of smaller tributary networks many millions of years to respond and demonstrate efficient coupling (Reinhardt et al. 2007).

The link between landslides' locations and migrating knickpoints suggests that slope failures are more likely to occur downstream of knickpoints. Migrating knickpoints trigger landslides, rockfall, talus accumulation and impose a spatial pattern on slope failures. The frequency of landslides per unit area will generally increases downstream toward the mouth, i.e., tends to be largest in zones dominated by a falling base-level. Such spatial pattern is very different compared to a landslide distribution pattern triggered by earthquakes.

After the knickpoint has undercut and destabilized the toe of the hillslope and the incisional wave has passed and migrated upstream, hillslope activity will decay and the slope system will start to stabilize. The greater concentration of landslides immediately downstream of a major knickpoint will gradually decrease, landslides will die out, and no major renewed slope failures are to be expected along the former landsliding area except due to tectonic and/or climatic events (Hovius et al. 1998).

12.4 Divide Migration

Divides are controlled by intrinsic factors as rock erodibility and tectonics and by extrinsic controls as precipitation and base-level. Divides become mobile when bounded by different channel gradients on either side and are stable when balanced between the magnitude and frequency of the tectonic activity compared to the available stream power (Mather 2000; Robl et al. 2017). An active motion of the drainage divide implies cross-divide contrasts of the erosion rates. A divide will migrate from the steeper side with the higher erosion to the side with lower erosion rates, relative higher base-level and lower slope gradients (Zondervan et al. 2020). Knickpoints activated by a lowering base-level may, over long time spans, migrate and result in piracy, especially if activated by a steep and short stream (Zaprowski et al. 2001; Figs. 12.4 and 12.5). Divide migration and piracy may change endorheic to exoreic conditions dominated by a lower base-level.

The impact of a base-level change is of high magnitude around a capture site, where from a wave of incision propagates spatially to adjust the captured area (Stokes et al. 2002). A significant impact will be the change in the captured valley shape. The transient signal in form of an incisional wave into the captured segment is part of a *drainage reorganization* (Yanites et al. 2013). Stored sediments in the captured tributary will start to be flushed and transported out and added to the channel, now dominating the entire drainage system but still graded for transporting low-sediment discharges as in the precapture stage.

Divide migration is part of the erosional struggle around a relief peak. Relief and nearby base-levels disappear with time (Fig. 12.6). When does the relief evolution reach its peak following the tectonic uplift stage, during the incisional cycle, triggered by base-level fall? We expect relief evolution to come to its ultimate stage during the initial incision period, while crustal elevation loss still keeps minimal, i.e., when low rates of erosion still typify the upland surfaces (Stock et al.

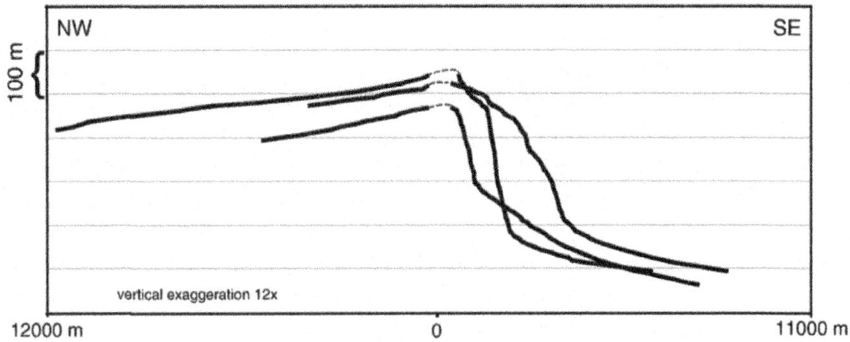

Fig. 12.5 Typical morphological contrasts between longitudinal profiles across an asymmetric structure with high (NW) and a low (SE) base-levels. The SE steep flank controls high competent channels of focused erosion. Repeated capturing results in a north-westward retreating divide (Redrawn from Prince et al. 2010)

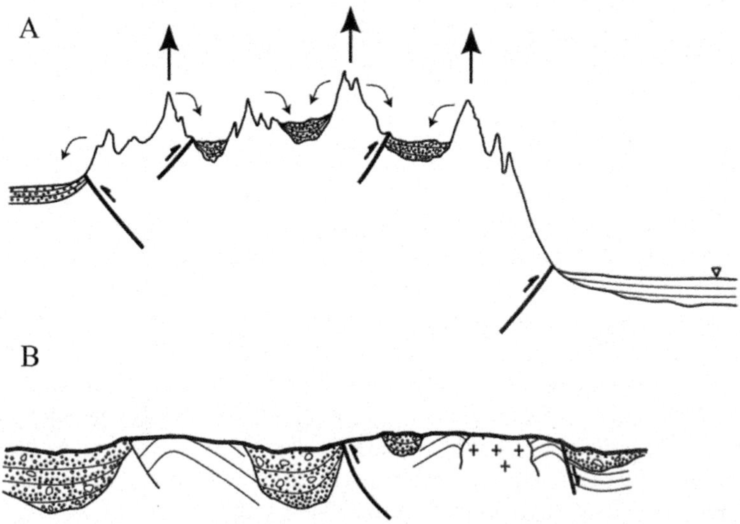

Fig. 12.6 **A** Highlands and lowlands in a mountain building evolution stage form closed inter-mountain base-levels at different altitudes connected to short local drainage systems. **B** Infilling the basins and eroding the mountains smooths the surface. The internal base-level pattern becomes gradually lost and a peneplain plateau surface domains (based on Liu-Zeng et al. 2008)

2004). At this initial stage of relief production, after the mountainous uplift event (base-level fall) and during the initial incision response, the drainage network at high altitudes still erodes at slow rates thereby giving way to a significant max topographic mountainous relief.

12.5 Entrapped Channel Morphology Under Rapid and Continuous Base-Level Fall

During the response to base-level fall, channel shape is expected to progress systematically through a continuum of sequential changes and adjustments which include the following stages suggested in conceptual models (Schumm et al. 1984; Simon and Hupp 1986; Elliott et al. 1999; Simon and Rinaldi 2006): (i) Most of the energy is directed to vertical incision and formation of a narrow and deeply incised channels with concentrated flow that, by positive feedback, increases stream power per unit width. This channel narrowing stage appears to be the dominating primary means by which fluvial systems initially respond to base-level fall (Doyle et al. 2003). V-shaped cross sections mark this stage of *erosional narrowing* typified by a high shear stress. (ii) The continual vertical incision induces slope failures. Landsliding becomes the primary process of the hillslope adjustment that initiates channel widening and energy dissipation by reducing the hydraulic depth and increasing relative roughness. (iii) The base-level stabilized and the time post-base-level drop becomes the main control. Widening gradually increases the stability. During this channel evolutional phase, with stages that may partly overlap, energy per unit water discharge (i.e., average cross-sectional shear stress) decreases with time (Simon 1992).

However, how does morphology react to an *extreme, rapid and, ongoing* base-level drop? The Dead Sea with its extreme mean annual level drop, exceeding 1 m (see Chap. 16), provides the conditions to record such channel shape evolution forced by an *extreme rapid and continuous base-level drop* (Bowman et al. 2010). Under such extreme conditions, the incised channel shows a trapezoidal, or v-shaped, gorge-like cross-sectional form with steep banks (Fig. 12.7), typified by tension cracks and failed slabs (Shapira 2007). The rapid lowering of the base-level entrapped the channels in continuous vertical incision, preventing the evolutionary progression toward a wide, graded and stable geometry which has been conceptualized in models. Under such severe and continuous base-level drop, no widening counteracted deepening to become the principal mechanism of dissipating the excess stream power (Carson 1984).

In addition to morphology, signatures of base-level fluctuations can be stored in the stratigraphic and textural record. To mention a few: *Stratigraphic disconformities* (Figs. 11.2 and 12.8) are signals of base-level fall and rise (Martin et al. 2009). An *upward textural coarsening* trend and an *increase in coarse grain size basinward* are often a criterion in identifying a lowering base-level. *Basinward shift of facies* may be as well a criterion. Base-level rise will result in a landward translation of depositional environments and in increasing the accommodation space. In many cases sequence boundaries are interpreted to reflect a hiatus developed during base-level fall (Shanley and McCabe 1994). *Planar erosional surfaces* form during a relatively slow sea level fall. *Incised valleys* preserved in the stratigraphic record indicate a relatively rapid base-level drop, and *Paleosols* reflect a pedogenic time interval of an unentrenched paleo-surface.

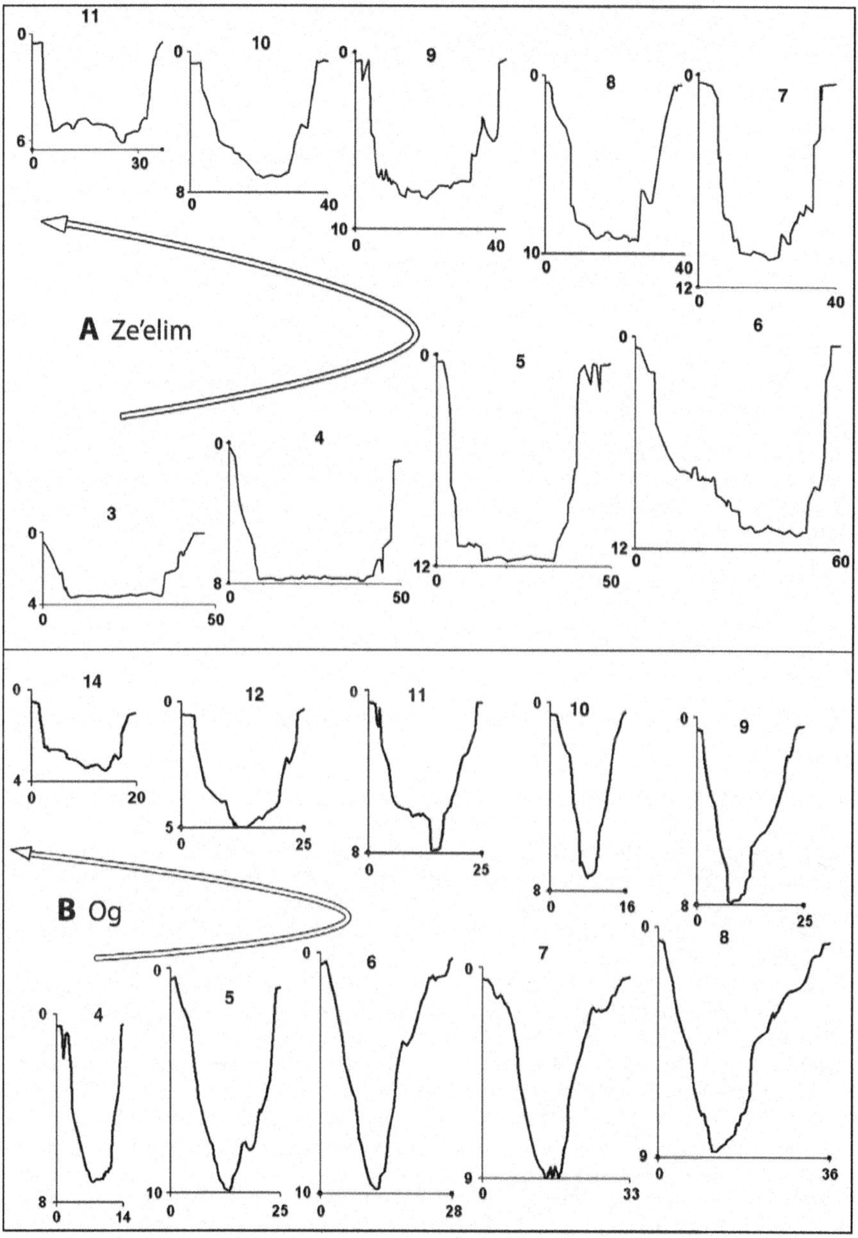

Fig. 12.7 Cross-sectional shapes in meter of N. Zeelim (**A**) and N. Og (**B**). Numbering is headward along 1–2 km from the outlet. Both channels are entrenched in lacustrine, silty-loam lake beds. V-shaped cross sections dominate the Og, quite different from the more trapezoidal Ze'elim sections. Based on field surveys. For location see Fig. 16.2 (following Shapira 2007)

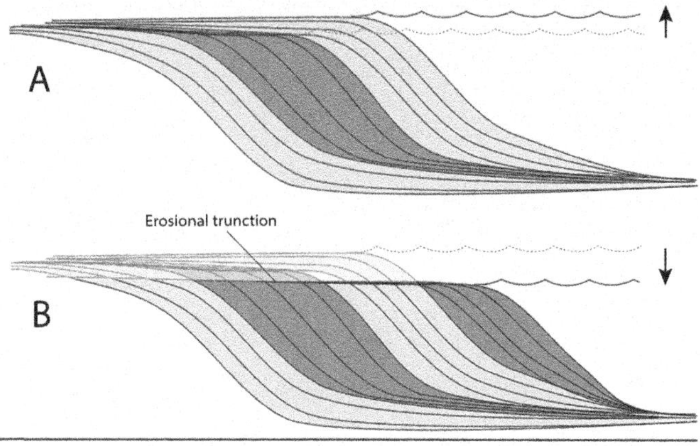

Fig. 12.8 Schematic longitudinal stratigraphic sections of a Gilbert-type delta, showing the link between foreset facies and base-level change. The cartoon depicts a sigmoidal package formed during base-level rise (**A**) and fluvial truncation (**B**) by incision during base-level fall (From Gobo et al. 2015)

References

Barnes JB, Densmore AL, Mukul M, Sinha R, Jain V, Tandon SK (2011) Interplay between faulting and base level in the development of Himalayan frontal fold topography. J Geophys Res: Earth Surf 116(F3)

Bevis M (2015) Sediment budgets indicate Pleistocene base level fall drives erosion in Minnesota's greater Blue Earth River basin. Doctoral dissertation, University of Minnesota

Bigi A, Hasbargen LE, Montanari A, Paola C (2006) Knickpoints and hillslope failures: interactions in a steady-state experimental landscape. Geol Soc Am Spec Pap 398:295

Bowman D, Svoray T, Devora S, Shapira I, Laronne JB (2010) Extreme rates of channel incision and shape evolution in response to a continuous, rapid base-level fall, the Dead Sea, Israel. Geomorphology 114(3):227–237

Brocard GY, Willenbring JK, Scatena FN, Johnson AH (2015) Effects of a tectonically-triggered wave of incision on riverine exports and soil mineralogy in the Luquillo Mountains of Puerto Rico. Appl Geochem 63:586–598

Carretier S, Lucazeau F (2005) How does alluvial sedimentation at range fronts modify the erosional dynamics of mountain catchments? Basin Res 17(3):361–381

Carson MA (1984) The meandering braided river threshold: a reappraisal. J Hydrol 73:315–334

Clark MK, Maheo G, Saleeby J, Farley KA (2005) The non-equilibrium landscape of the southern Sierra Nevada, California. GSA Today 15(9)

Clark MK, Royden LH, Whipple KX, Burchfiel BC, Zhang X, Tang W (2006) Use of a regional, relict landscape to measure vertical deformation of the eastern Tibetan Plateau. J Geophys Res: Earth Surf 111(F3)

Crosby BT (2006) The transient response of bedrock river networks to sudden base-level fall. Doctoral dissertation, Massachusetts Institute of Technology

Crosby BT, Whipple KX (2006) Knickpoint initiation and distribution within fluvial networks: 236 waterfalls in the Waipaoa River, North Island, New Zealand. Geomorphology 82(1–2):16–23

Dapples F, Oswald D, Raetzo H, Lardelli T, Zwahlen P (2003) New records of holocene landslide activity in the western and eastern Swiss alps: implication of climate and vegetation changes. Eclogae Geol Helv 96:1–9

Densmore AL, Dawers NH, Gupta S, Guidon R (2005) What sets topographic relief in extensional footwalls? Geology 33(6):453–456

Densmore AL, Ellis MA, Anderson RS (1998) Landsliding and the evolution of normal-fault-bounded mountains. J Geophys Res: Solid Earth 103(B7):15203–15219

DiBiase RA, Whipple KX, Lamb MP, Heimsath AM (2015) The role of waterfalls and knickzones in controlling the style and pace of landscape adjustment in the western San Gabriel Mountains, California. Bulletin 127(3–4):539–559

Doyle MW, Stanley EH, Harbor JM (2003) Channel adjustments following two dam removals in Wisconsin. Water Resour Res 39(1)

Elliott JG, Gellis AC, Aby CB (1999) Evolution of arroyos: incised channels of the Southwestern United States. In: Darby SE, Simon A (eds) Incised river channel. Wiley, Chichester, UK, pp 153–185

Gallen SF, Wegmann KW, Frankel KL, Hughes S, Lewis RQ, Lyons N, Witt AC (2011) Hillslope response to knickpoint migration in the Southern Appalachians: implications for the evolution of post-orogenic landscapes. Earth Surf Proc Land 36(9):1254–1267

Gardner TW (1983) Experimental study of knickpoint and longitudinal profile evolution in cohesive, homogeneous material. Geol Soc Am Bull 94(5):664–672

Gobo K, Ghinassi M, Nemec W (2015) Gilbert-type deltas recording short-term base-level changes: delta-brink morphodynamics and related foreset facies. Sedimentology 62(7):1923–1949

Harkins N, Kirby E, Heimsath A, Robinson R, Reiser U (2007) Transient fluvial incision in the headwaters of the Yellow River, northeastern Tibet, China. J Geophys Res: Earth Surf 112(F)

Hovius N, Stark CP, Tutton MA, Abbott LD (1998) Landslide-driven drainage network evolution in a pre-steady-state mountain belt: Finisterre Mountains, Papua New Guinea. Geology 26(12):1071–1074

Korup O, Schlunegger F (2007) Bedrock landsliding, river incision and transience of geomorphic hillslope-channel coupling: evidence from inner gorges in the Swiss Alps. J Geophys Res: Earth Surf 112(F3)

Liu-Zeng J, Tapponnier P, Gaudemer Y, Ding L (2008) Quantifying landscape differences across the Tibetan plateau: implications for topographic relief evolution. J Geophys Res: Earth Surf 113(F4)

Martin J, Paola C, Abreu V, Neal J, Sheets B (2009) Sequence stratigraphyof experimental strata under known conditions of differential subsidence and variable base-level. AAPG Bull 93(4):503–533

Mather AE (2000) Adjustment of a drainage network to capture induced base-level change. Geomorphology 34:271–289

Mather AE, Stokes M, Whitfield E (2017) River terraces and alluvial fans: the case for an integrated quaternary fluvial archive. Quat Sci Rev 166:74–90

Merritts D, Vincent KR (1989) Geomorphic response of coastal streams to low, intermediate, and high rates of uplift, medocino triple junction region, Northern California. Geol Soc Am Bull 101(11):1373–1388

Merritts DJ, Vincent KR, Wohl EE (1994) Long river profiles, tectonism, and eustasy: a guide to interpreting fluvial terraces. J Geophys Res: Solid Earth 99(B7):14031–14050

Morell KD, Kirby E, Fisher DM, Van Soest M (2012) Geomorphic and exhumational response of the Central American Volcanic Arc to Cocos Ridge subduction. J Geophys Res: Solid Earth 117

Mudd SM, Furbish DJ (2007) Responses of soil-mantled hillslopes to transient channel incision rates. J Geophys Res: Earth Surf 112(F3)

Muto T, Steel RJ (2001) Autostepping during the transgressive growth of deltas: results from flume experiments. Geology 29(9):771–774

Ouimet WB, Whipple KX, Granger DE (2009) Beyond threshold hillslopes: channel adjustment to base-level fall in tectonically active mountain ranges. Geology 37(7):579–582

Prince PS, Spotila JA (2013) Evidence of transient topographic disequilibrium in a landward passive margin river system: knickpoints and paleo-landscapes of the New River basin, Southern Appalachians. Earth Surf Proc Land 38(14):1685–1699

Prince PS, Spotila JA, Henika WS (2010) New physical evidence of the role of stream capture in active retreat of the Blue Ridge escarpment, Southern Appalachians. Geomorphology 123(3–4):305–319

Reinhardt LJ, Bishop P, Hoey TB, Dempster TJ, Sanderson DCW (2007) Quantification of the transient response to base-level fall in a small mountain catchment: Sierra Nevada, Ssouthern Spain. J Geophys Res: Earth Surf 112(F3)

Robl J, Heberer B, Prasicek G, Neubauer F, Hergarten S (2017) The topography of a continental indenter: the interplay between crustal deformation, erosion and base-level changes in the Eastern Southern Alps. J Geophys Res Earth Surf 122(1):310–334

Schumm SA, Harvey MD, Watson CC (1984) Incised channels: initation, evolution, dynamics and control. Water Resources Publication

Shanley KW, McCabe PJ (1994) Perspectives on the sequence stratigraphy of continental strata. AAPG Bull 78(4):544–568

Shapira I (2007) Cross-sectional shape and bank processes in micro-canyons incised into lacustrine sediments in the North-Western Dead Sea shore. MA Thesis, Ben-Gurion University of the Negev, Beer Sheva, Israel. In Hebrew with English abstract

Simon A (1992) Energy, time, and channel evolution in catastrophically disturbed fluvial systems. Geomorphology 5:345–372

Simon A, Hupp CR (1986) Channel evolution in modified Tennessee channels. In: Proceedings of the 4th Federal interagency sedimentation conference. US Government Printing Office, Washington DC, pp 71–82

Simon A, Rinaldi M (2006) Disturbance, stream incision and channel evolution: the roles of excess transport capacity and boundary materials in controlling channel response. Geomorphology 79:361–383

Sklar L, Dietrich WE (1998) River longitudinal profiles and bedrock incision models: stream power and the influence of sediment supply. Geophys Monograph-Am Geophys Union 107:237–260

Stock GM, Anderson RS, Finkel RC (2004) Pace of landscape evolution in the Sierra Nevada, California, revealed by cosmogenic dating of cave sediments. Geology 32(3):193–196

Stokes M, Mather AE, Harvey AM (2002) Quantification of river-capture-induced base-level changes and landscape development, Sorbas Basin, SE Spain. Geol Soc Londn Special Publ 191(1):23–35

Weissel JK, Seidl MA (1997) Influence of rock strength properties on escarpment retreat across passive continental margins. Geology 25(7):631–634

Weissel JK, Seidl MA (1998) Inland propagation of erosional escarpments and river profile evolution across the Southeast Australian passive continental margin. Geophys Monograph-Am Geophys Union 107:189–206

Whipple KX, Dibiase RA, Crosby BT (2013) Bedrock rivers. In: Treatise on geomorphology. Elsevier, pp 550–573

Whittaker AC, Cowie PA, Attal M, Tucker GE, Roberts GP (2007) Bedrock channel adjustment to tectonic forcing: implications for predicting river incision rates. Geology 35(2):103–106

Wobus CW, Hodges KV, Whipple KX (2003) Has focused denudation sustained active thrusting at the Himalayan topographic front? Geology 31(10):861–864

Yanites BJ, Ehlers TA, Becker JK, Schnellmann M, Heuberger S (2013) High magnitude and rapid incision from river capture: Rhine River, Switzerland. J Geophys Res Earth Surf 118(2):1060–1084

Zaprowski BJ, Evenson EB, Pazzaglia FJ, Epstein JB (2001) Knickzone propagation in the Black Hills and Northern high plains: a different perspective on the late Cenozoic exhumation of the Laramide Rocky Mountains. Geology 29(6):547–550

Zhang H, Zhang P, Prush V, Zheng D, Zheng W, Wang W, Ren Z (2017) Tectonic geomorphology of the Qilian Shan in the northeastern Tibetan plateau: insights into the plateau formation processes. Tectonophysics 706:103–115

Zondervan JR, Stoke M, Boulton SJ, Telfer MW, Mather AE (2020) Rock strength and structural controls on fluvial erodibility: Implications for drainage divide mobility in a collisional mountain belt. Earth Planet Sci Lett 538:116221

The Transient-Equilibrium Test 13

Abstract

Landscapes are generally classified as being either in steady state or in transient conditions. Steady state, equilibrium or graded state refer to a status where the channel had adjusted to the base-level and to other climatic, lithologic and tectonic controlling factors. The base-level is one of the factors that conditions equilibrium and may initiate degradation or aggradation in order to return to equilibrium. A graded channel shows neither aggrading nor incising major trends and has accomplished transmitting the effect of base-level fall upstream along the profile. A transient morphology marks disequilibrium and has yet to adjust to the base-level and to the other controlling conditions. Being in a transient state means that the channel is neither in energy equilibrium nor in a topographic steady state, but is progressing toward it.

Keywords

Steady state/Equilibrium • Transient state • Relict morphology • Relaxation time

13.1 Steady State Criterions

Examining the base-level from the viewpoint of equilibrium/disequilibrium may improve our understanding its impact. Leopold and Bull (1979), Mackin (1948), Morisawa (1985) and Howard (1982) characterized a graded stream neither by strong trends of deposition nor by entrenchment. Equilibrium refers to insensitivity of the channel gradient to high-frequency changes of the hydraulic regime (Howard 1982). Mackin (1948) defined a graded stream as one where, over a period of years, slope is delicately adjusted to provide, with the available discharge and the prevailing channel characteristics, just the energy required for transporting the load supplied from the drainage basin. In order to form a stable equilibrium, the longitudinal profile, including its slope, flow velocity, depth, width, roughness, pattern and bed morphology, has to be mutually adjusted to provide the minimum

D. Bowman, *Base-level Impact*,
https://doi.org/10.1007/978-3-031-24994-5_13

power and maximum efficiency that is only necessary to transport the load supplied from the drainage basin (Leopold and Bull 1979).

Whipple (2001) suggested that channel elevation change depends in general on the imbalance between uplift and erosion and that *steady-state orogeny* is where erosion balances rock uplift. The incision of a river during steady state perfectly counterbalances rock uplift rate so that elevations along the channel profile are unchanging in time. A river crossing a zone of active uplift with incisional capability that matches the uplift rate would have reached conditions of equilibrium (Whittaker et al. 2007), i.e., an *erosional steady state* when erosion rate equals rock uplift rate at all locations. In equilibrium state, the bed sediment influx equals the transport rate at any given point in the drainage network and erosion balances the base-level lowering so that the topography remains preserved in time.

Quaternary climate fluctuation was often too rapid to allow for true steady-state conditions to recover. A significant morphologic adjustment may not have time to occur, i.e., the *equilibrium time* (Van Heijst and Postma 2001) needed for the fluvial system to regain equilibrium from the moment that it was disturbed by base-level change was unavailable. Equilibrium needs maintenance of the velocities necessary to shear and transport the coarser debris. This necessitates downstream changes in the form of the cross section and decrease in grain size. A *Graded status* implies absence of large-magnitude base-level changes, or large-magnitude cyclical changes of the hydraulic regime. Channels in steady state are characterized by monotonic relationship between channel steepness and erosion rate (DiBiase et al. 2015).

A main criterion for equilibrium is whether, over a period of time, the channel remains stable. The condition of a channel at any given time is however not necessarily a reliable indicator of its long-term status because bed morphology can change quickly in response to short term events as episodic climatic or fluvial changes. The stability or movement of drainage divides can serve as a marker of a steady or disequilibrium state (Willett et al. 2014). To examine the horizontal motion or the stability of drainage divides, a straightforward way of mapping is needed. Equilibrium is thus represented by an almost similar morphology along the channel without abrupt changes. Constant energy dissipation per unit area of the channel is expected with minimum energy dissipation across the network.

Profile concavity is often consistent with streams in a graded state and transition to a concave-up profile can be interpreted as return to steady-state conditions. Graded valley beds have often, close to their base-level, a cover of alluvium. Flat bottom morphology of a valley indicates that it is under stable and graded morphological conditions rather than under an effective degradational regime. Rivers approach equilibrium asymptotically, i.e., the wave of equilibration propagates upstream fast at the start and the rate of change decreases exponentially (Howard 1965). Close to the base-level, steady-state erosion rates are firstly reached. Further upstream steady state will be exceeded later. The farthest regions close to the divide have the most delayed response.

13.2 Transient Activity

Incision or deposition following base-level changes are adjustments of the network toward grade after being pushed by external forcing to disequilibrium (Whipple 2001). Non-graded morphology is in a *transient state*, i.e., it has to adjust to new controls, including base-level conditions. In a transient landscape, the area is progressively adjusting but may still show relict morphology of the former, past environmental conditions. If the relaxation time of the drainage system, i.e., the time required for a new equilibrium to be attained, is much longer than the frequency of the climatic-tectonic oscillations, the drainage system will be unable to return to grade.

Deviations from the power law slope–area relationship may serve as evidence for a transient evolution (Tucker and Whipple 2002; Stock et al. 2005). A period that transforms landscapes into significantly high relief typifies the transient stage (Burbank and Pinter 1999). The hydraulic and slope adjustments, as described in the theory of minimum variance (Leopold and Langbein 1962), are part of the self-regulating feedback mechanism that operates during the transient period in the direction toward providing the minimum velocity required for transporting the load supplied from upstream.

Profile analysis can serve as a primary test to estimate a graded or a transient status. A major fall of the base-level will cause downcutting and readjustment of the stream gradients during the transient period until a new graded profile is reestablished (Schumm 1979). Smaller streams and tributaries are steeper than larger ones and maintain in their transient status the power necessary to incise. Typical features of an active transient fluvial regime include as well: migrating water divides; gorge morphology and strong coupling of channels to the valley sides with input of debris to the channel and knickpoints that do not correlate with changes in rock mass strength. All serve as typical non-equilibrium markers of the transient regime.

Steady state is achieved after the transient period, when vertical incision rates match base-level lowering. In a steady-state landscape, river incision rates balance everywhere the rock uplift rate (Whipple and Tucker 2002) and streams have accomplished transmitting the effect of the base-level fall upstream along the profile. Large-scale flattened relief and lowered divides are related to the termination of the transient period upon reaching the final equilibrium state, when the wave of erosion reached the channel headwater and knickpoints have been eliminated (Crosby and Whipple 2006). The return to equilibrium is achieved by successive upstream wave trains of knickpoints: first, wave trains that induce the release of the alluvial cover, followed by second wave trains that drive the bedrock incision (Loget and Van Den Driessche 2009).

Catchments which have reached topographic steady state do not display grain-size signals (Whittaker et al. 2007), i.e., coarse grain size injected somewhere into the longitudinal profile. Lack of knickpoints may as well serve as an indication for certain stability, although the absence of knickpoints in longitudinal profiles does not necessarily indicate that the network has reached steady state. Hillslope processes have a relaxation time to adjust that may be longer than the time needed for channels to adjust to base-level fall. Therefore, hillslopes extend the duration of landscape transience (Reinhardt et al. 2007) and delay achieving steady state.

References

Burbank DW, Pinter N (1999) Landscape evolution: the interactions of tectonics and surface processes. Basin Res 11(1):1–6

Crosby BT, Whipple KX (2006) Knickpoint initiation and distribution within fluvial networks: 236 waterfalls in the Waipaoa River, North Island, New Zealand. Geomorphology 82(1–2):16–38

DiBiase RA, Whipple KX, Lamb MP, Heimsath AM (2015) The role of waterfalls and knickzones in controlling the style and pace of landscape adjustment in the western San Gabriel Mountains, California. Bulletin 127(3–4):539–559

Howard AD (1965) Geomorphological systems-equilibrium and dynamics. Am J Sci 263:302–312

Howard AD (1982) Equilibrium and time scales in geomorphology: application to sand-bed alluvial streams. Earth Surf Proc Land 7(4):303–325

Leopold LB, Bull WB (1979) Base-level, aggradation and grade. Proc Am Philos Soc 123:168–202

Leopold LB, Langbein WB (1962) The concept of entropy in landscape evolution, vol 500. US Government Printing Office

Loget N, Van Den Driessche J (2009) Wave train model for knickpoint migration. Geomorphology 106(3–4):376–382

Mackin J (1948) Concept of the graded river. Geol Soc Am Bull 59(5):463–512

Morisawa M (1985) Rivers: form and process. Geomorphology texts, vol 7

Reinhardt LJ, Bishop P, Hoey TB, Dempster TJ, Sanderson DCW (2007) Quantification of the transient response to base-level fall in a small mountain catchment: Sierra Nevada, southern Spain. J Geophys Res Earth Surf 112:F3

Schumm SA (1979) Geomorphic thresholds: the concept and its applications. Trans Inst Br Geogr 485–515

Stock GM, Anderson RS, Finkel RC (2005) Rates of erosion and topographic evolution of the Sierra Nevada, California, inferred from cosmogenic ^{26}Al and ^{10}Be concentrations. Earth Surf Process Land J Br Geomorphol Res Group 30(8):985–1006

Tucker GE, Whipple KX (2002) Topographic outcomes predicted by stream erosion models: sensitivity analysis and intermodel comparison. J Geophys Res Solid Earth 107(B9):ETG-1

Van Heijst MW, Postma G (2001) Fluvial response to sea-level changes: a quantitative analogue experimental approach. Basin Res 13(3):269–292

Whipple KX (2001) Fluvial landscape response time: how plausible is steady-state denudation? Am J Sci 301(4–5):313–325

Whipple KX, Tucker GE (2002) Implications of sediment-flux-dependent river incision models for landscape evolution. J Geophys Res Solid Earth 107(B2):ETG-3

Whittaker AC, Cowie PA, Attal M, Tucker GE, Roberts G (2007) Characterizing the transient response of rivers crossing active normal faults: new field observations from Italy. Basin Res 19:529–556

Willett SD, McCoy SW, Perron JT, Goren L, Chen CY (2014) Dynamic reorganization of river basins. Science 343:6175

14 Tributary Junctions

Abstract

Trunk–tributary contrast in drainage area often leads to a sharp difference in their incision capability which commonly leads to formation of a knickpoint at the tributary junction, i.e., at the confluence of the tributary with the main stem. Tributaries adjust to the mainstream as to a local base-level and are directly uncontrolled by the ultimate base-level downstream. Tributary junctions may steepen and become a knickpoint which separates the main trunk stream from the often "hanging" conditions of the tributary valley. Hanging conditions limit the connectivity and impede upstream transmission of incision signals and thus delay reaching full equilibrium of the drainage network.

Keywords

Hanging valley • Tributary

The difference in drainage area between a main stem and his tributaries may lead to a sharp difference in their incision capability. At the confluence with the trunk stream the incisional difference cause elevation differences of the channel beds. The trunk channel functions as a lowered base-level, triggering a knickpoint to propagate up the tributary. Tributaries are thus uncontrolled directly by the ultimate base-level. The main trunk functions as their effective local base-level. Catastrophic floods in trunk streams sharply lower them as base-level of the tributaries which incise in response. As small tributaries cannot incise at rates equal to the incision in the mainstream, *hanging valleys*, i.e., elevated tributary basins form above the trunk stream, segregated by an oversteepened reach (Crosby et al. 2007; Wobus et al. 2006) that forms a knickpoint or knickzone. Hanging valleys are steepening toward the deeper incised mainstream (Fig. 14.1; Crosby and Whipple 2006) and become insulated from it (Whipple et al. 1999; Fig. 14.2). Many hanging valleys have a glacial origin, i.e., an evolution that started from an abrupt glacial incision along a main trunk that became disconnected by waterfalls and step-pool morphology from the small tributaries (Valla et al. 2010). The *tributary*

D. Bowman, *Base-level Impact*,
https://doi.org/10.1007/978-3-031-24994-5_14

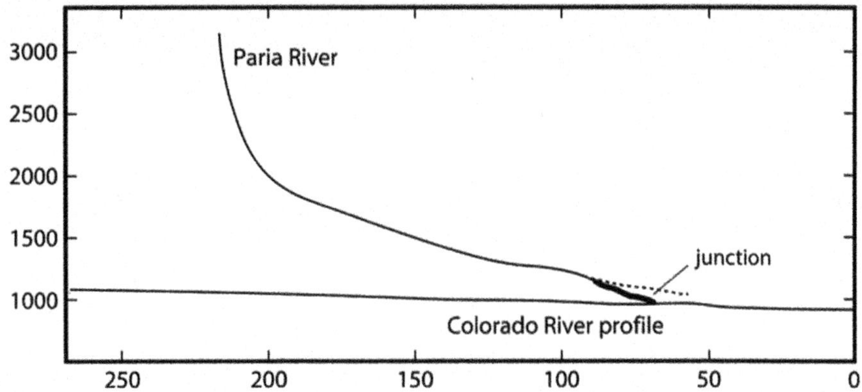

Fig. 14.1 Setting of a tributary junction in a drainage network: The oversteepened, lower segment of the Paria River tributary composes a *tributary junction*, marked by either a distinct knickpoint or a broad knickzone when entering the Colorado River (Modified after Cook et al. 2009)

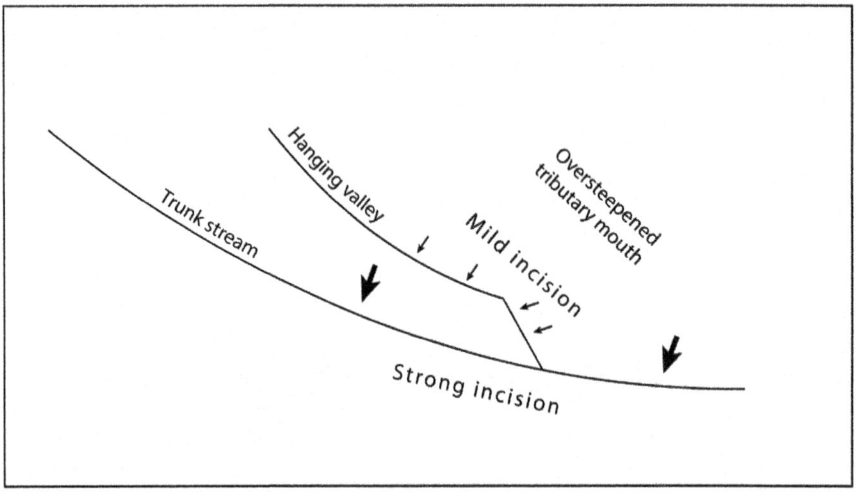

Fig. 14.2 Strong incision along the trunk stream relative to minor incision along the tributary valley form the *hanging valley* morphology. The communication of erosional signals up the hanging valley is limited

junction, i.e., the confluence of two different incision rates becomes a knickpoint or knickzone, i.e., a local base-level, through rapids, a waterfall or a gorge that are keeping the tributary elevated above the mainstream (Clark et al. 2006; Fig. 14.3). Hanging valleys thus comprise the upper-floor B in an incised drainage network, (Reed 1981). The trunk–tributary contrasts are intrinsic and form without any exogenic impacts (Hayakawa and Oguchi 2014).

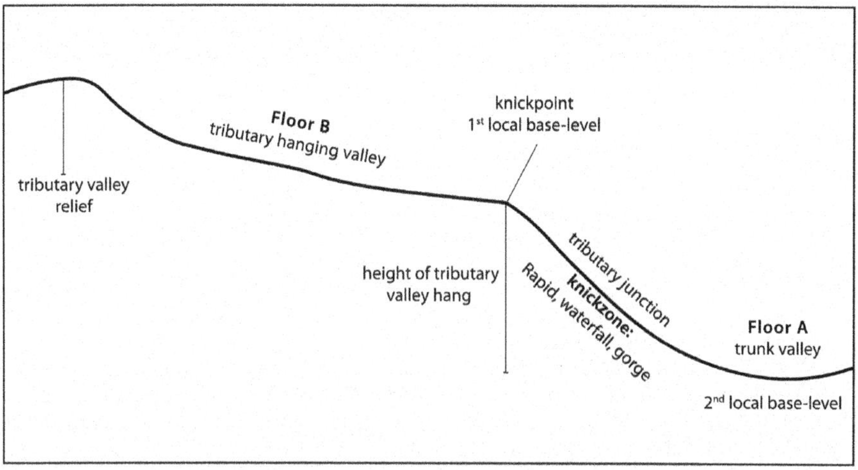

Fig. 14.3 Post-glacial-fluvial tributary junction environment, showing a two-floor morphology, including two local base-levels insulating the hanging valley from the trunk local base-level. The gradients at tributary mouths, along the knickzone, are significantly oversteepened and often pass a threshold value beyond which erosional efficiency declines. The tributary low local relief (hundreds of m) can be found on the high-elevation (thousands m) of the hanging valley (floor B) which has not yet responded to any deep, active ultimate base-level

When small tributaries enter large main stem rivers, as a result of their significant smaller drainage area, the tributary junction knickpoint may stay anchored for extended periods of time. The mainstream may, under such conditions, continue incising very rapidly, or the sediment flux from the tributary may become very small resulting with a tributary junction that will continue to oversteepen and further stagnate the propagation of the incision from the trunk river into the tributary. However, there may evolve *temporary* hanging valleys that eventually recover and equilibrate to the main stem, as for example, when high rates of decay of the incision signal, prevents deep incision as it propagates upstream along the main stem. High-elevation and low-relief surfaces in the hanging valleys may gradually evolve into an equilibrated relief by grading toward the junction knickpoint.

The typical smaller drainage area of hanging valleys suggests that there might be a threshold magnitude of drainage area required to provide enough incision capability to prevent generation of hanging valleys. In a study of hanging valleys in the Coast Ranges of Taiwan, Wobus et al. (2006) recognized that most hanging tributaries have a trunk-to-tributary drainage area ratio greater than ~10:1 which supports the observation that main stem drainage area must be significantly greater than the tributary drainage area in order to create significant hanging conditions. Tributaries located near the outlet of the main trunk stream feel the fastest incision rate of the trunk river, being sourced by the entire drainage basin and having the greatest potential to create hanging valleys. Upper-basin tributaries experience a lower incision of the main trunk. Distributing the fall of the main stem, as an incising base-level, over a longer time period, tributary junctions have a greater probability of keeping pace with the incision and thus lowering the likelihood of creating hanging valleys.

References

Clark MK, Royden LH, Whipple KX, Burchfiel BC, Zhang X, Tang W (2006) Use of a regional, relict landscape to measure vertical deformation of the eastern Tibetan Plateau. J Geophys Res: Earth Surf 111(F3)

Cook KL, Whipple KX, Heimsath AM, Hanks TC (2009) Rapid incision of the Colorado River in Glen Canyon–insights from channel profiles, local incision rates, and modeling of lithologic controls. Earth Surf Proc Land 34(7):994–1010

Crosby BT, Whipple KX (2006) Knickpoint initiation and distribution within fluvial networks: 236 waterfalls in the Waipaoa River, North Island, New Zealand. Geomorphology 82(1–2):16–38

Crosby BT, Whipple KX, Gasparini NM, Wobus CW (2007) Formation of fluvial hanging valleys: theory and simulation. J Geophys Res: Earth Surf 112:F03S10

Hayakawa YS, Oguchi T (2014) Spatial correspondence of knickzones and stream confluences along bedrock rivers in Japan: implications for hydraulic formation of knickzones. Geogr Ann Ser B 96(1):9–19

Reed JC Jr (1981) Disequilibrium profile of the Potomac River near Washington, DC–A result of lowered base level or quaternary tectonics along the fall line? Geology 9(10):445–450

Valla PG, van der Beek PA, Lague D (2010) Fluvial incision into bedrock: Insights from morphometric analysis and numerical modeling of gorges incising glacial hanging valleys (Western Alps, France). J Geophys Res 115

Whipple KX, Kirby E, Brocklehurst SH (1999) Geomorphic limits to climate-induced increases in topographic relief. Nature 401(6748):39–43

Wobus CW, Crosby BT, Whipple KX (2006) Hanging valleys in fluvial systems: controls on occurrence and implications for landscape evolution. J Geophys Res: Earth Surf 111(F2)

Small-Scale Networks and Man-Made Structures 15

Abstract

In addition to natural channel networks, base-level rules dominate as well small-scale phenomena such as rills and are relevant to man-made structures too. Headcut of rills that form and extend upslope are activated by local, small-scale base-level lowering. Retreat of gullies is a larger phenomenon, but still makes a relative small-scale demonstration of knickpoint propagation. Check-dams are raised local base-levels that promote deposition and may function as small waterfalls with scour below, whereas dams are significant man-made knickpoints. Removal of dams is lowering a local base-level back to the natural channel level by the height of the impoundment. After the removal and once the reservoir water has drained, a narrow channel starts to migrate upstream through the reservoir deposits, analogous to alluvial channels responding to base-level fall. A gravel extraction site in a channel bed is a man-made lowered local base-level. On its upstream end, incision typically evolves as a retreating knickpoint. Dewatering reservoirs means their conversion from a sediment sink to a sediment source. The magnitude and rate of reservoir sediment erosion, following dam removal, starts initially large and rapid, but diminishes substantially. The water released from dams before removal, is relatively free of sediments and capable of downcutting. Notably, downstream of dams, incision is initiated as a common response operating from upside downcurrent, opposite to the natural upstream migration of base-level impacts.

Keywords

Rills • Gullies • Check-dams • Gravel mining • Headcutting

15.1 Small-Scale Networks

Rills are ephemeral concentrated flow paths, centimeters in cross section, which function as both sediment source and transport system in upland areas and on slopes. Experiments (Gordon et al. 2011) showed that rill headcuts form and

D. Bowman, *Base-level Impact*,
https://doi.org/10.1007/978-3-031-24994-5_15

extend upstream activated by small-scale base-level lowering and that changes in the cumulative sediment flux and erosion depth matches the base-level drops. The impacts of a lowering base-level are quickly communicated to all portions of the evolving rill network.

Gullies compose the first significant delivery system with width and depth dimensions of up to a few meters (Poesen et al. 2003). The gullies are steep walled, sharply incised with heads of an erosional scarp form demonstrating a small-scale knickpoint propagation. The gully headcut retreat is controlled by washing, crumbling, slumping and caving, assisted by tensional cracks development, soil toppling, collapsing pipes and plunge pool undermining.

The headcut of a gully is located between the incised part and the upstream nourishing drainage area that is controlled by rills. The retreat of the gully headcut is linked with its drainage area which is a surrogate for surface runoff volume and rainfall depth, including basin properties such as bedrock, soil and vegetation cover. The evolution of the gully headcut is enhanced by failures at wetter zones within the profile (Roloff et al. 1981). Plunge pool erosion and an impinging jet with subsurface flow, slumping, piping and seepage erosion play an important role in the headcut migration.

The mean rate of headcut retreat was found <1.5 m/y for gullies developing in sandy, marly and clay deposits. Vanmaercke et al. (2016) reported gully headcut median retreat rate of 0.89 m/y, significantly correlated to the runoff contributing drainage area and a rainy day factor. Oostwoud Wijdenes and Bryan (2001) reported of annual gully head retreat rates from 0.8 to 3 m, with extremes of 15 m/y. The riverbed lowering as the base-level drops causes increase in the gully headcut erosion and in landslide frequency (Sidorchuk 2006; Bogen et al. 1994; Nichols et al. 2016).

15.2 Man-Made Structures

Instream gravel mining—sediment mining is often carried out extensively along alluvial rivers (Rinaldi 2003). Extraction of sediments lowers the channel bed within the boundaries of the gravel mine and induces channel incision that will migrate as a knickpoint (Fig. 15.1) many kilometers upstream from the mine (Kondolf 1997). Gravel mining in channels thus means **a.** *forming a local lower base-level* which traps much of the incoming bedload sediments passing downstream and **b.** *upstream migration of bed incision* that starts from the upper tip of the extraction site. A headcut of 4 m migrated hundreds of m upstream. Headcuts propagated upstream even for kilometers in the main river and up tributaries (Stevens et al. 1990). Once gravel extraction stops, the channel may continue to adjust to the removal of the bed sediments for years. Mining of bed material and trapping in the extraction site much of the bedload passing downstream reduces bedload supply downstream of the mining site.

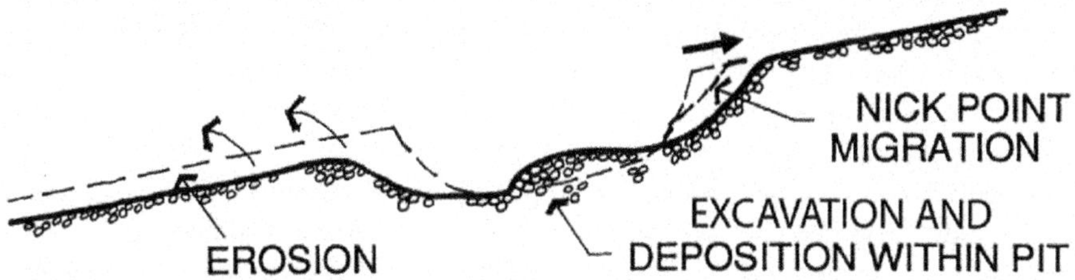

Fig. 15.1 An extraction gravel pit as a local lowering base-level with headcutting eroding upstream (following Kondolf 1997)

Check-dams are raised, local base-levels, constructed with rocks, boulders or gravel, built across gullies and first-order channels, promoting local deposition by reducing the channel gradient while functioning as a waterfall with scour below (Shit et al. 2013; Fig. 15.2) Agriculturally the major function of check-dams was to reduce water velocity and to trap sediments in badland areas (Singh and Dubey 2002) .

Small check-dams are intended to control the undercutting of valley side slopes by preventing channel incision through trapping sediments. The dams are only short-term effective measures for sediment control; i.e., it can be effective for reducing sediment yield only temporarily (Zhao et al. 2017). When check-dams fill with sediment, the flat surface of the sediment wedge provide arable land for agriculture. Check-dams trap and retain upstream a great part of the sediment as a wedge that decreases in thickness upstream (Fig. 15.2). The width of the deposited

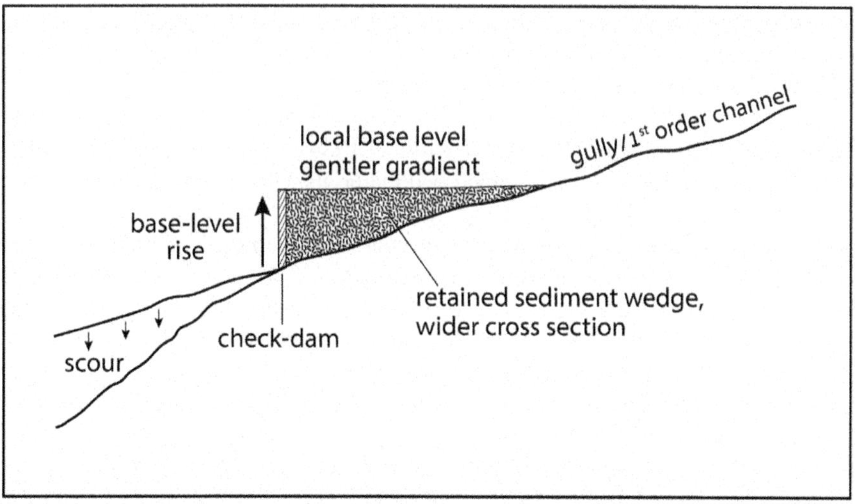

Fig. 15.2 Rising base-level effect of check-dams. Demonstrated is the sediment accumulation due to the dam barrier and the decrease in channel gradient. Downstream the channel incises

sediment behind the dam becomes greater than the width of the natural channel, thus decreasing the flow depth and the velocity (Leopold 1992). The intersection of the water surface in the dam with the channel bed upstream functions as a raised local base-level, the effect of which is felt only a short distance of several hundred meters upstream.

Check-dams were built by Nabatean people of the Negev Desert, Israel, (third century BC–first century AC) along wadi beds at certain distances apart. Agriculturally, the major function of check-dams was to trap sediments in badland areas. The infilled areas served the purpose for growing crops. By filling with trapped sediments, the dams formed a new gradient, smaller than that of the original channel. The new gradient varies between thirty and sixty percent of that which characterized the original channel bed and was still sufficient to transport all the input of sediments (Leopold 1992).

Landslide dams are formed following natural rock and debris avalanches. The mechanisms triggering dam-forming landslides are excessive rainfall, snowmelt and earthquakes. The landslide dams do not join the permanent "dam family" and fail shortly after formation.

Dams are built to supply power, reduce flood hazard, improve navigation or impound water for irrigation and urban water supply. Most dams are small, i.e., <10 m tall. The reservoir behind the dam is functioning as a base-level, trapping both water and sediments. The water released from the reservoir, being relatively free of sediments, can be termed "hungry water" (Kondolf 1997), possessing stream power in excess of the sediment load and tending to erode the channel bed. Immediately downstream of dams, incision is initiating as a common response and the channel demonstrates a drastic shrinkage, often to a narrow single thread. These channel adjustments are typically greatest during the first high flows after dam closure and immediately downstream of the dam, decreasing over time and distance from the dam (Williams and Wolman 1984). Channel narrowing has been greatest below reservoirs that are large enough to contain the river's largest floodwaters, allowing encroachment of vegetation.

The downstream effects of dams may extend hundreds kms and may continue up to 30 years or more (Williams and Wolman 1984). Incision below dams is most pronounced in rivers with fine-grained bed materials. Channel erosion below dams is frequently accompanied by change in the particle size on the bed following winnowing that leaves an armor layer of coarse lag deposit that limits the depth of incision (Dietrich et al. 1989). To compensate for the lack of sediment supply downstream of the dams and prevent habitat and infrastructure damage by channel incision, sediments may be artificially added to channels (Kondolf et al. 2014). It is possible to pass sediments and thereby maintain some continuity of sediment transport through the system. This is the recommended design for sediment management.

The degrading processes downstream of dams uniquely operates from upside downcurrent, opposite to the natural upstream trend of migrating base-level impacts. The migration period of the front of the degraded zone means a lag for the far downstream reaches. The downstream migrating degradation front may meet an upstream prograding incision trend triggered by a falling base-level.

Dam removal—Dams are removed for many reasons, commonly because their loss of function has rendered them unsafe or economically non-viable. Many reservoirs become filled with sediment to the point of being useless and many other dams are close to this endpoint. Another key motivation to remove dams is to rehabilitate rivers by restoring physical and ecological attributes and functions (Pohl 2002). Notably, from the longitudinal channel profile point of view, the dam is a man-made knickpoint, i.e., a local base-level (Graf 2006). Removing a dam is just as lowering a local base-level by the height of the impoundment, i.e., a knickpoint destruction. The dam height controls the amount of water and sediment in storage and determines the possible local base-level fall.

The removal strategy, i.e., a sudden removal of the dam or a gradual removal, will define the rate of the base-level fall. The reaction time may thus be short or prolonged—hours to days—following a sudden removal, or months to years during a phased removal. A series of multiple, small knickpoints tend to form and migrate upstream and gradually release sediment from the impoundment during phased removals (Foley et al. 2017). Major et al. (2017) summed up the geomorphic response to dam removal: after the reservoir water has been drained, a narrow channel starts to form at the dam site and migrates upstream crossing the reservoir (Fig. 15.3). The deepening results in erosional narrowing (Ferrer-Boix et al. 2014) followed subsequently by widening. A knickpoint or knickzone will migrate upstream through the impounded sediment until the channel gradient approaches the predam channel slope. The change in the channel gradient through the reservoir reach is always toward the predam profile (Fig. 15.4).

Surveys showed that the magnitude and rate of reservoir erosion start large. Rivers respond rapidly after dam removal and transport the sediments eroded from the reservoirs quickly, trending toward their predam morphology (Foley et al. 2017). The rapid dam removal may cause exceptionally high sediment transport rates that are in orders of magnitude greater than the background rates, but diminishes substantially and the knickpoint simultaneously decreases in height and becomes less distinct (Major et al. 2012). Reported nickpoint migration rates varied from as much as 200 m/hr in the first hours following breaching to less than 2 m/d by the end of the first year.

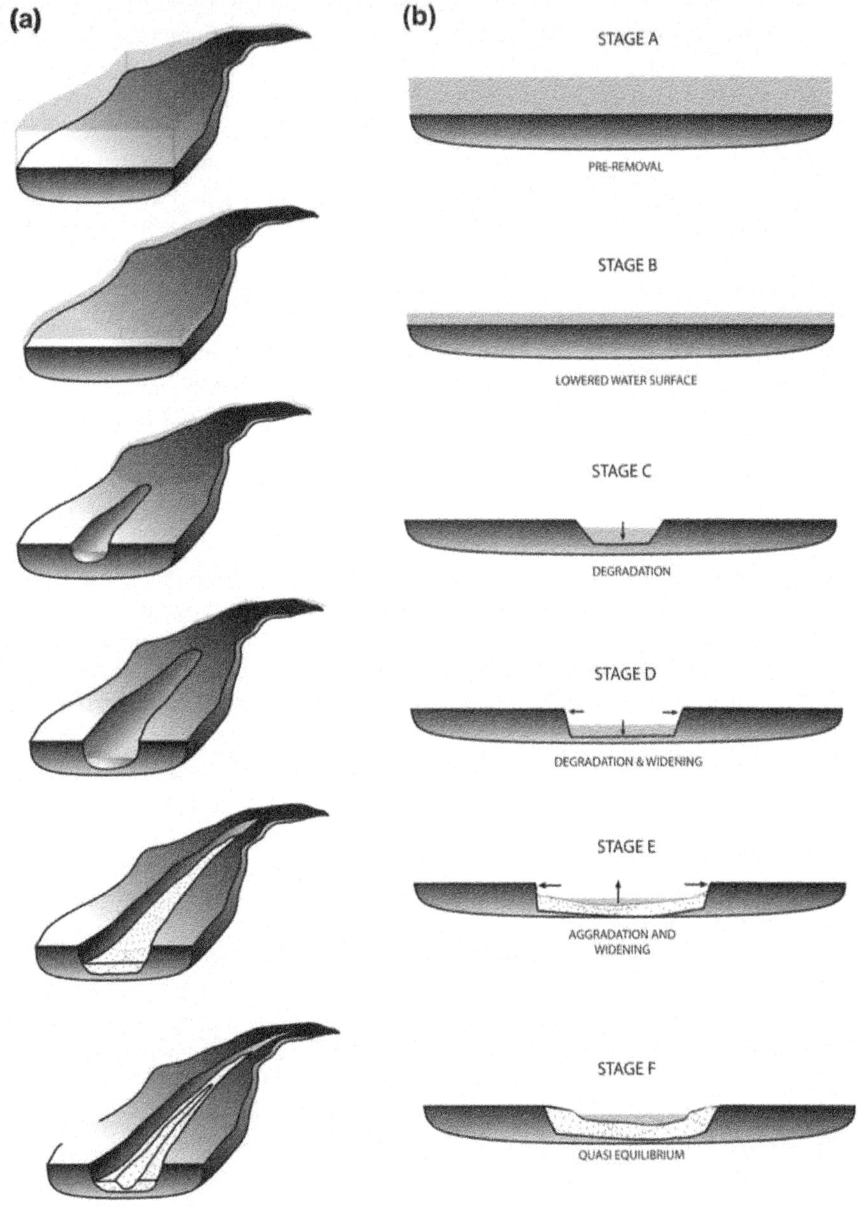

Fig. 15.3 Channel evolution following removal of a low head dam representing a lowering base-level effect. The cross sections demonstrate through time: incision, widening and aggradation of sediments derived from upstream erosion (From Doyle et al. 2003)

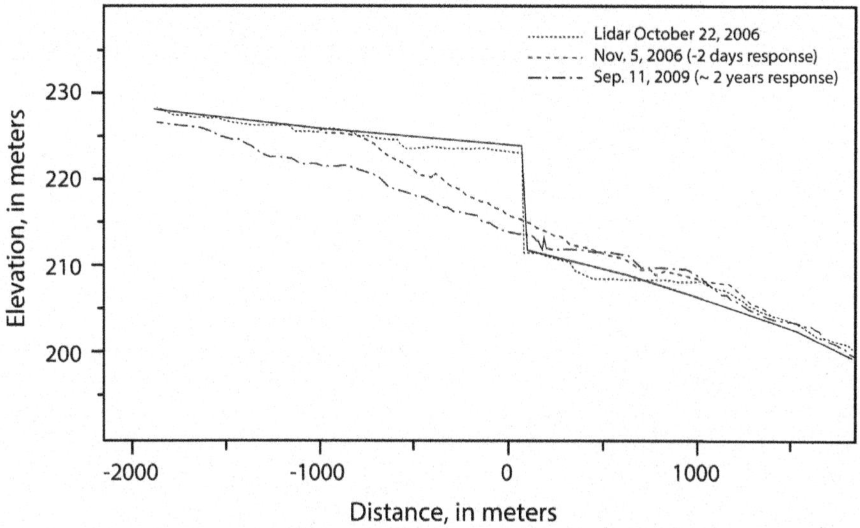

Fig. 15.4 Measured profiles in the reservoir reach of the Marmot Dam USA demonstrating, right after the dam removal, the large and rapid removal of sediments stored behind the dam and the diminishing rate of erosion and the deposition downstream (Redrawn from Major et al. 2012)

Following dam removal, the material trapped in the reservoir becomes the principal source of the sediment transported downstream (Doyle et al. 2003). When the sediment supply from the removed dam exceeds the river's transport capacity, deposition will occur (Fig. 15.5) within a few kilometers downstream of dam site which may increase the channel gradient and width, causing filling of pools. Dam removal finally tends to reestablish an upstream and downstream connectivity, i.e., restores the movement of water, material and organisms between the upstream and downstream river reach.

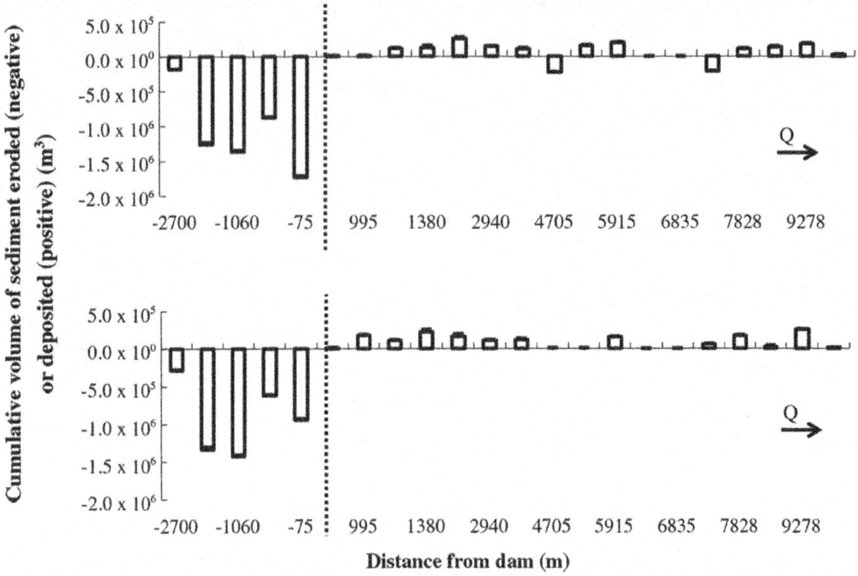

Fig. 15.5 Response to Barlin Dam failure in Taiwan 2007, in two different periods during a Typhoon. Cumulative net volumes of erosion and deposition are shown in the reservoir and along the downstream channel. The vertical dashed line represents the location of the dam. Flow (*Q*) direction is to the right. The magnitude of downstream deposition, following the dam failure, is small relative to the volume eroded from the reservoir. The discrepancy suggests that much of the sediment eroded from the reservoir was transported downstream beyond the study reach. Erosion moved sediments into the reservoir from the far-upstream, demonstrating an upstream transient base-level erosional effect. The changing sediment distribution downstream to the reservoir indicates sediment pulses (From Tullos and Wang 2014)

References

Bogen J, Berg H, Sandersen F (1994) The contribution of gully erosion to the sediment budget of the River Leira. IAHS Publ-Ser Proc Rep-Intern Assoc Hydrol Sci 224:307–316

Dietrich WE, Kirchner JW, Ikeda H, Iseya F (1989) Sediment supply and the development of the coarse surface layer in gravel-bedded rivers. Nature 340(6230):215–217

Doyle MW, Stanley EH, Harbor JM (2003) Channel adjustments following two dam removals in Wisconsin. Water Resour Res 39

Ferrer-Boix C, Martín-Vide JP, Parker G (2014) Channel evolution after dam removal in a poorly sorted sediment mixture: experiments and numerical model. Water Resour Res 50(11):8997–9091

Foley MM, Bellmore JR, O'Connor JE, Duda JJ, East AE, Grant GE, Wilcox AC et al (2017) Dam removal: listening in. Water Resour Res 53(7):5229–5246

Gordon LM, Bennett SJ, Wells RR (2011) Evolution of rill networks on soil-mantled experimental landscapes driven by rainfall and base-level adjustments. Landform Anal 17:57–63

Graf WL (2006) Downstream hydrologic and geomorphic effects of large dams on American rivers. Geomorphology 79(3–4):336–360

Kondolf GM (1997) Hungry water: effects of dams and gravel mining on river channels. Environ Manage 21(4):533–551

Kondolf GM, Gao Y, Annandale GW, Morris GL, Jiang E, Zhang J, Yang CT et al (2014) Sustainable sediment management in reservoirs and regulated rivers: experiences from five continents. Earth's Future 2(5):256–280

Leopold LB (1992) Base-level rise: gradient of deposition. Israel J Earth Sci 41:57–64

Major JJ, East AE, O'Connor JE, Grant GE, Wilcox AC, Magirl CS, Tullos DD et al (2017) Geomorphic responses to dam removal in the United States—a two-decade perspective In: Tsutsumi D, Laronne JB (eds) Gravel-bed rivers. Wiley, Hoboken, pp 355–383

Major JJ, O'Connor JE, Podolak CJ, Keith MK, Grant GE, Spicer KR, Wilcock PR et al (2012) Geomorphic response of the Sandy River, Oregon to removal of Marmot Dam. Reston, VA. US Geological Survey, Professional Paper 1792, p 64

Nichols MH, Nearing M, Hernandez M, Polyakov VO (2016) Monitoring channel head erosion processes in response to an artificially induced abrupt base-level change using time-lapse photography. Geomorphology 265:107–116

Oostwoud Wijdenes DJ, Bryan R (2001) Gully-head erosion processes on a semi-arid valley floor in Kenya: a case study into temporal variation and sediment budgeting. Earth Surf Proc Land 26(9):911–933

Poesen J, Nachtergaele J, Verstraeten G, Valentin C (2003) Gully erosion and environmental change: importance and research needs. Catena 50(2–4):91–133

Pohl MM (2002) Bringing down our dams: trends in American dam removal rationales. J Am Water Resour Assoc 38:1511–1519

Rinaldi M (2003) Recent channel adjustments in alluvial rivers of Tuscany, Central Italy. Earth Surf Proc Land 28(6):587–608

Roloff G, Bradford JM, Scrivner CL (1981) Gully development in the deep loess hills region of Central Missouri. Soil Sci Soc Am J 45(1):119–123

Shit PK, Bhunia GS, Maiti R (2013) Assessing the performance of check dams to control rill-gully erosion: small catchment scale study. Int J Curr Res 5(04):899–906

Sidorchuk A (2006) Stages in gully evolution and self-organized criticality. Earth Surf Processes Landforms: J Br Geomorphol Res Group 31(11):1329–1344

Singh S, Dubey A (2002) Gully erosion and management methods and applications (a field manual). New Academic Publishers, Delhi, p 248

Stevens MA, Urbonas B, Tucker LS (1990) Public–private cooperation protects river. APWA Reporter 25:7–27

Tullos D, Wang HW (2014) Morphological responses and sediment processes following a typhoon-induced dam failure, Dahan River, Taiwan. Earth Surf Processes Landforms 39(2):245–258

Vanmaercke M, Poesen J, Van Mele B, Demuzere M, Bruynseels A, Golosov V, Yermolaev O et al (2016) How fast do gully headcuts retreat? Earth Sci Rev 154:336–355

Williams GP, Wolman MG (1984) Downstream effects of dams on alluvial rivers. US Geol Surv Prof Pap 1286

Zhao G, Kondolf GM, Mu X, Han M, He Z, Rubin Z, Sun W et al (2017) Sediment yield reduction associated with land use changes and check dams in a catchment of the Loess Plateau, China. Catena 148:126–137

The Dead Sea Area as a Field Laboratory—A Regional Approach

16

Abstract

The Dead Sea area, located within the African Rift Valley, occupies the deepest depression on earth. Its hydrological regime is typified by long dry intervals, high-intensity short rain events, limited-area rain cells and ephemeral streams with flash floods of peaky hydrographs. High flow velocities attain 4–6 m/s along typically steep (> 1°) channels in an area dominated by soft erodible sediments. Lake Lisan, the Late-Pleistocene precursor of the Dead Sea, occupied the area at about 70–15 ka BP. The Dead Sea followed Lake Lisan as a hypersaline terminal base-level. Recently, the Dead Sea is typified by its rapid accelerating drop, reaching rates of over 1 m/year. This dramatic, continuous, human-induced base-level fall, as well as the soft lithology, makes the Dead Sea area an exceptional natural field laboratory for studying, in real time, the fluvial response to an extreme ongoing base-level drop. The typical divergence of river terraces downstream and the convexity of the channel profiles as well as their steepness are controlled in the Dead Sea area by the accelerating base-level fall and the control of the steep clinoform structure of the fan deltas. The gap between the recurrence interval of floods and the continuous rapid base-level drop causes channels to become temporally isolated from the rapid falling base-level at their front. The deeply incised, steep and sinuous reach of the Jordan River, entering the Dead Sea from the north, generally fits Schumm's conceptual model of sinuosity evolution by base-level drop. In the newly exposed Jordan delta, sinuosity increased from ~ 1.05 to 1.35. Along the western Dead Sea shore and Nahal (Wadi in Hebrew) Arava that enters the Dead Sea from south, there was no enhanced thalweg sinuosity recorded beyond 1.1–1.2. The hydraulic geometry of channels demonstrates initiation of widening by forming trapezoidal cross sections. Additional widening seems to be retarded by the continuous, rapid base-level drop. Channel incision rates in the Dead Sea area range from 0.3 to about 0.8 m/year^{-1}, which are global extreme rates. The channels are continuously in ungraded conditions, i.e., without the stability required for an equilibrium to be attained. Although being located at the lowest site on

D. Bowman, *Base-level Impact*,
https://doi.org/10.1007/978-3-031-24994-5_16

earth, hundreds of meters below the rift shoulders and below the global mean sea level, the Dead Sea proves, in many cases, hydraulically meaningless as an active controlling terminal base-level, demonstrating thereby the importance of the local base-level type as dominating the drainage network.

Keywords

Lake Lisan • The Dead Sea retreat • Divergence and Convergence of river terraces • Sinuosity • Incision rates • Connectivity • Headward entrenchment • Exposed lake bed

16.1 Setting

The Dead Sea area occupies the lowest elevation on the Earth surface. It is located along the *Dead Sea transform* within the *African Rift System*, a plate boundary that separates the Arabian plate from the Sinai subplate (Fig. 16.1). The first major subsidence of the *Dead Sea pull-apart basin* converted it to an inland local base-level that, from early Miocene, drained westward to the Mediterranean through a fluvial corridor (Bar and Zilberman 2016). It became later a regional inland base-level composed of several sub-basins. Its function as an inland basin is inferred from its sequence of lacustrine and evaporite sediments, more than 10 km thick (Quennell 1958; Garfunkel 2014).

The Dead Sea rift area is bounded on the east and west by tectonic escarpments. The area is surrounded to the west by the Jerusalem and Hebron mountains, 800–1000 m above sea level (Fig. 16.2, no. 7, 8) with mean annual precipitation of 500–600 mm/year^{-1} which decrease to 50 mm/year^{-1} over the Dead Sea. To the east (Fig. 16.2, no. 10, 11), the Dead Sea area is bounded by the Jordanian plateau and the Moab and Edom Mountains, 1700 m asl with mean annual rainfall exceeding 300 mm year^{-1}. The Dead Sea area is an hyper-arid rain shadow desert lying about 430 m below the mean sea level (Fig. 16.3) at the lowest rift and world altitude, with annual potential evaporation of about 2 m, mean annual temperature of 25 °C and maximum daily temperature exceeding 40 °C.

Hydrological data for the Dead Sea area is based mainly on the field studies of Cohen (2005), Greenbaum et al. (2006) and Zituni et al. (2021). The hydrological regime is typified by long dry intervals, high-intensity short rain events as well as limited-area rain cells and ephemeral flash floods of short, peaky hydrographs. Maximum flood events per year in the period 1996–2003 were nine with two dry years. Flood activity lasts for very short periods, leaving the drainage system dry more than 99% of the year. Nevertheless, floods may occur during the autumn, winter and spring. The high flow velocities along the steep channels that slope 0° 49′–1° 25′ attain 4–6 m/s. Bedload fluxes are among the highest ever recorded; 2–10 kg/m/year and a few tens of kg/m/year are not infrequent. However, in spite of the imposing work done during short flood events, the long-term mean annual load yield is low due to the long recurrence intervals of the flood (Greenbaum et al. 2006).

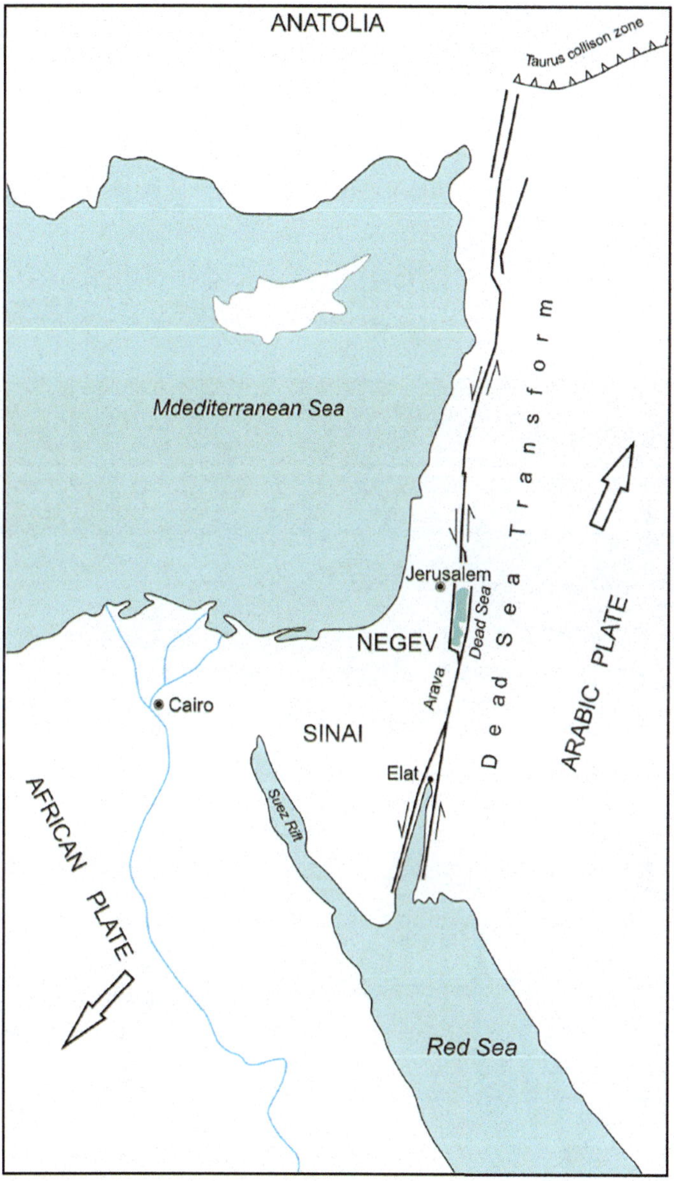

Fig. 16.1 The Dead Sea transform marks the plate boundary which separates the Arabian Plate from the African plate and Sinai

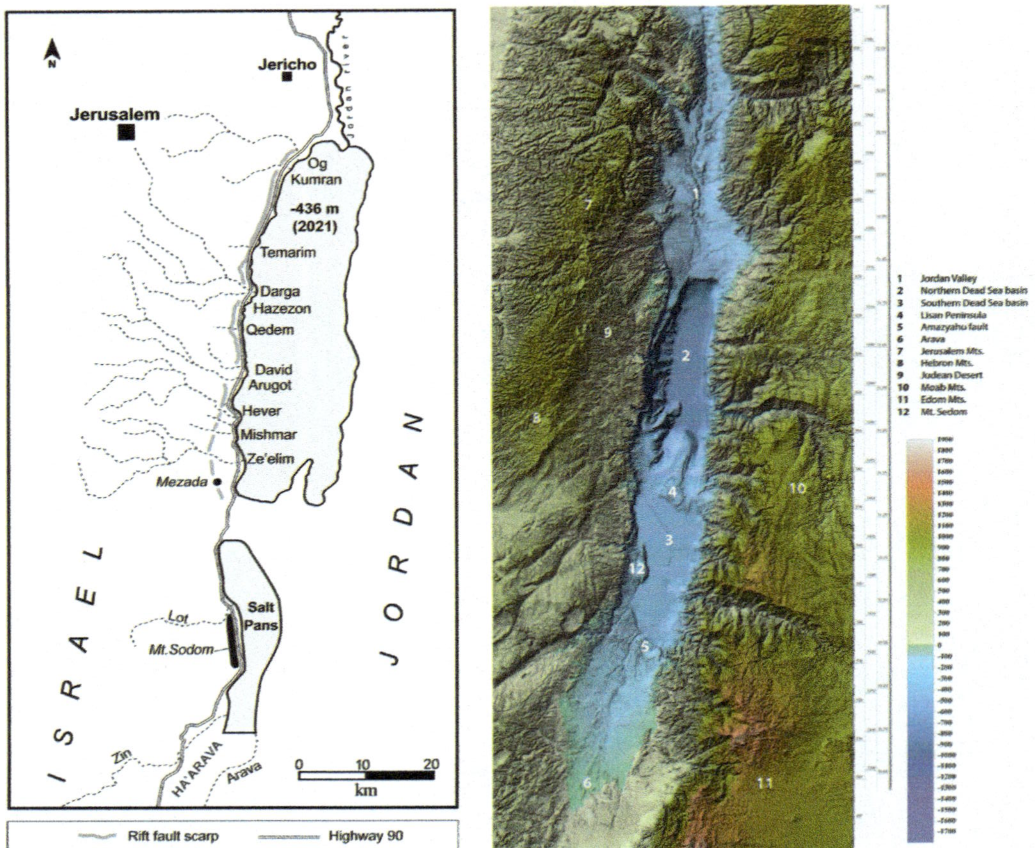

Fig. 16.2 Left—location map. The scarp is shown along the lake. The location of the channels in the southern Dead Sea basin is not shown. On the right—digital hypsometric colored and shaded relief map of the Dead Sea area and its surroundings by J. K. Hall and R. Calvo

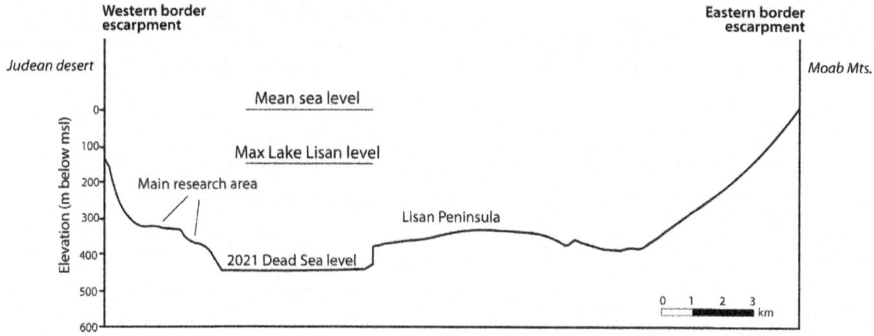

Fig. 16.3 General E–W cross section through the Dead Sea Rift north of Mezada demonstrating the lake levels in relation to the Mediterranean sea level

The marginal rift escarpment which separates the Dead Sea area (2 in Fig. 16.2) from the Judean desert (9 in Fig. 16.2) is an abrupt knickzone composed of waterfalls and cascades, cut into Cenomanian–Turonian hard rocks. At the base of the rift escarpment, along the western side of the Dead Sea, successions of alluvial fans radiate from canyons to be deposited as a veneer across the Dead Sea shore. The morphology of the area was summarized by Bowman (1971, 1977, 1978, 1988, 1990, 1997). The Lake Lisan sedimentary sequence exposed in the area is basically composed of aragonite laminae, gypsum, halite, silt and clay, often with carbonate coarse sand, interbedded with an alluvial sequence of poorly stratified and sorted pebbles and cobbles to large boulders with fine-bedded sand and granules beds. Absence of exposed hard rocks and dominance of soft erodible sediments without channel-bed armor typify the high erodibility of the Dead Sea area and make it an ideal site to demonstrate the base-level impact. Studies along the western side of the Dead Sea are also relevant to the eastern coast of the lake, where incision rates may be of importance for future infrastructure design (Briaud et al. 2001).

16.2 Lakes as Fluctuating Base-Levels

Lake Lisan is the Late-Pleistocene precursor of the Dead Sea, dated 70–15 ka BP (Kaufman 1971; Schramm et al. 2000; Haase-Schramm et al. 2004). The lake reached its highest stand at about ~25 ka (Bartov et al. 2002) consistent with the Heinrich Event 3 which indicates a positive water balance in the Near East. Its high level corresponds to the last Glacial Wurm period (marine isotope stages 2–4) and at the transition to the Holocene it dropped dramatically. The Lisan lake did not occupy only the Dead Sea depression but extended from the Sea of Galilee in the north to the south of the recent Dead Sea (Fig. 16.4). Following its setting, the Lisan lake was a terminal, inland base-level. The highest stand of Lake Lisan, ~160/150 m below mean sea level, was morphologically indicated in the rift, south to the Dead Sea, by relict nearshore bars and abandoned shorelines, composed of unconsolidated coastal gravel (Bowman and Gross 1992). This was further confirmed by Klinger et al. (2003) and Matmon et al. (2003) through the use of Optically Stimulated Luminescence datings proving these shorelines to be Upper Pleistocene. The maximum high stand of Lake Lisan was 285 m higher than the present (2021) Dead Sea level, 435 m below the mean sea level (Fig. 16.3).

During the drop of Lake Lisan, shore terraces were carved and well preserved on the rift border escarpments (Bowman 1971; Ghazleh and Kempe 2009;

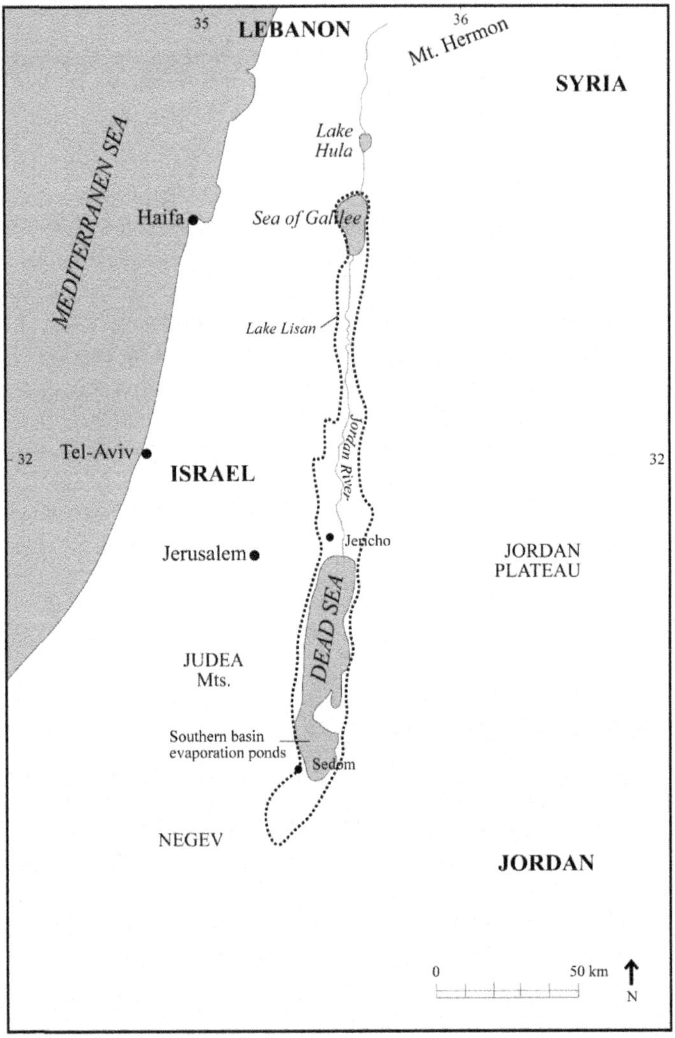

Fig. 16.4 The Dead Sea and Lake Lisan terminal lakes along the rift. Lake Lisan maximum extent at ca. 25 ka BP is shown by a dashed line (modified after Greenbaum et al. 2006)

Fig. 16.5). These relict shorelines are recessional terraces that record a series of stillstands during the falling Lisan lake. The terraces range in elevation from about 150 m bsl, which is the maximum highstand elevation of Lake Lisan, to almost the modern Dead Sea lake level. Ages of tufa at the elevation of 180 and 370 m bsl were determined to be 24 and 23 ka BP respectively (Neev 1964; Neev and Emery 1967), pointing to an abrupt drop of Lake Lisan level in a short period of 1000 year, equal to a falling rate of about 20 cm/year, which is significantly less than that of the recent falling rate of the Dead Sea level. The fall of the Lisan level could be the result of either a tectonic subsidence (Neev and Emery 1967) or change in the hydroclimatic conditions in the large watershed of the lake as part of Levant region that gained from the expansion of the European ice sheet, which had deflected westerly storm tracks southward to the Mediterranean and the climatic change decreased the cyclogenesis and moisture delivery to the Levant (Torfstein et al. 2013).

Valleys that drain toward the rift, often named *canyons*, have been flooded by Lake Lisan and were infilled with lacustrine sediments to be later evacuated after Lake Lisan fall. Some residuals of this Lisan lacustrine fill are still preserved along the canyons' slopes.

Dead Sea **-** The Dead Sea Lake followed Lake Lisan since 12,000/14,000 YBP as a Holocene hyper-saline terminal lake, one of the most hyper-saline lakes in the world (33.7%), an order of magnitude higher than seawater. The lake is 330 m deep, with the lake floor at ~ 725 m below mean sea level, located at the deepest spot in the rift and on earth. Recently, the Dead Sea lake became restricted to its deep northern basin, while the shallow southern basin (Fig. 16.2, no. 3; Fig. 16.4) would be dry, but is essentially covered by artificial evaporation ponds. The lake is recharged by freshwater from the Jordan and the Arava Rivers, by smaller rivers and by saline springs. The catchment of the Dead Sea is ~43,000 km^2.

The Dead Sea is unique in its rapid accelerating level drop over a short time scale of tens of years. The average rate of the lake level lowering during the period 1981–1988 was 0.8 m/year. The rate of lowering increased sharply from 1990 onward, reaching over 1 m/year. At 31/1/2021, the Dead Sea level was 435.31 m below mean sea level (Fig. 16.6). The negative water balance during the nineteenth century was mainly caused by the decrease in precipitation, whereas during the second half of the twentieth century and the beginning of the twenty-first century, it has been due to human interference which included building a series of dams on the larger Rivers in Syria and Jordan such as Yarmouk, Zarka and Mujib, building the Naharayim hydroelectric dam in 1927 and a regulating dam on the Jordan River in 1932, operation of the National Water Carrier by Israel in 1964 through which water was diverted from the Sea of Galilee and operating the Ghor Canal by Jordan. In 1982, the Dead Sea level dropped to 402 bsl, and the lake was separated into two basins. The Southern Basin of the Dead Sea was transformed into an increasing number of evaporation ponds both in Israel and in Jordan. About 40 cm/year of the lake level drop (40%) results from the significant amount of

Fig. 16.5 Raised beach terraces composed of wave-cut treads and risers indicating the final drop of Lake Lisan base-level. The terraces are cut by gullies (photo 2006 by Ofek Aerial Photography, Netanya, Israel)

water pumped from the Northern Basin of the Dead Sea toward the evaporation ponds.

The former higher lake levels of the Dead Sea are recorded by raised morphological coastal features as relict beaches, wave-cut benches and gravel bars, as well as by tar and driftwood deposits. The Dead Sea rapid accelerating level drop is 2–3 scales of magnitude faster than any other published rates (Schumm and Chorley 1983), and it is assumed that the lake level will continue to drop in the coming decades (Elias et al. 2011; Lensky and Dente 2015). Following the Dead Sea level fall, the recently exposed lake bed provides an opportunity to follow formation of new drainage networks that show dendricity in their initial channel evolution (Bowman et al. 2011; Fig. 16.7).

The dramatic, continuous, human-induced base-level fall, on time scales of several tens of years, makes the Dead Sea a unique field study area. The Dead Sea is in

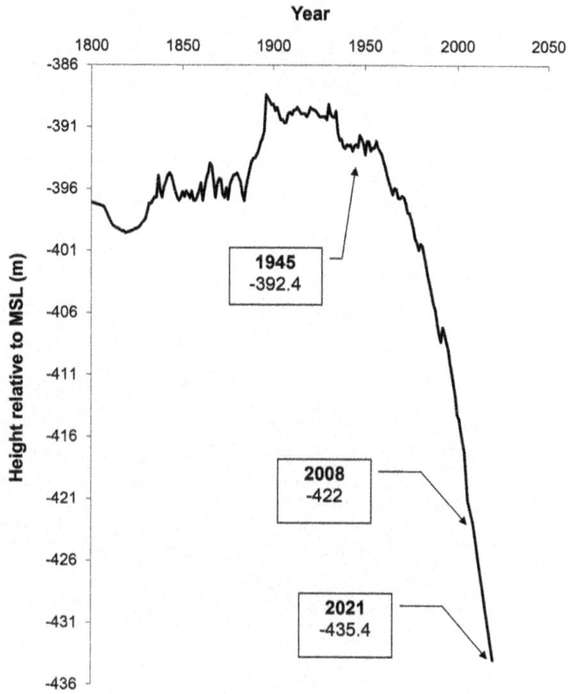

Fig. 16.6 Fluctuations of the Dead Sea level since 1850. Data based on: Masterman (1913), Ashbel (1965), Klein (1986), the Dead Sea Potash Works, the Israel Hydrological Service and the Government Water Authority

a continuous, non-relaxation stage that does not reach final stable conditions. The drainage networks are continuously adjusting to a rapid base-level fall and thus are trapped in constant ungraded conditions. Furthermore, the dominance of unconsolidated fan-delta gravelly deposits results in a high erodibility of the nearshore zone and contributes to the fast migration of the base-level impact headward. The rapid base-level fall and the soft lithology provide an exceptional natural field laboratory for studying in real time the response to an extreme ongoing base-level forcing.

16.3 Incision

North–south highway 90, along the Israeli Dead Sea shoreline, crosses different morphologies and demonstrates different responses to a rapid falling base-level. Upon crossing an alluvial fan, as for example, N. Ze'elim (Fig. 16.2) the prograding fan sediments bury the road that without maintenance would be completely buried after every flood event. Thus, although not far from the receding lake, the road is dominated by *prograding* fans, demonstrating *disconnecti*vity from the falling base-level. In another case, highway 90 crosses N. Darga, a few hundred m from the shoreline. The highway concrete protection forms a knickpoint that functions as a local, man-made base-level which prevents upstream progradation of the Dead Sea fall impact (Fig. 16.8). The same can be observed in N. Hazezon, N. Arugot and N. David (Fig. 16.9). However, with time, following upstream

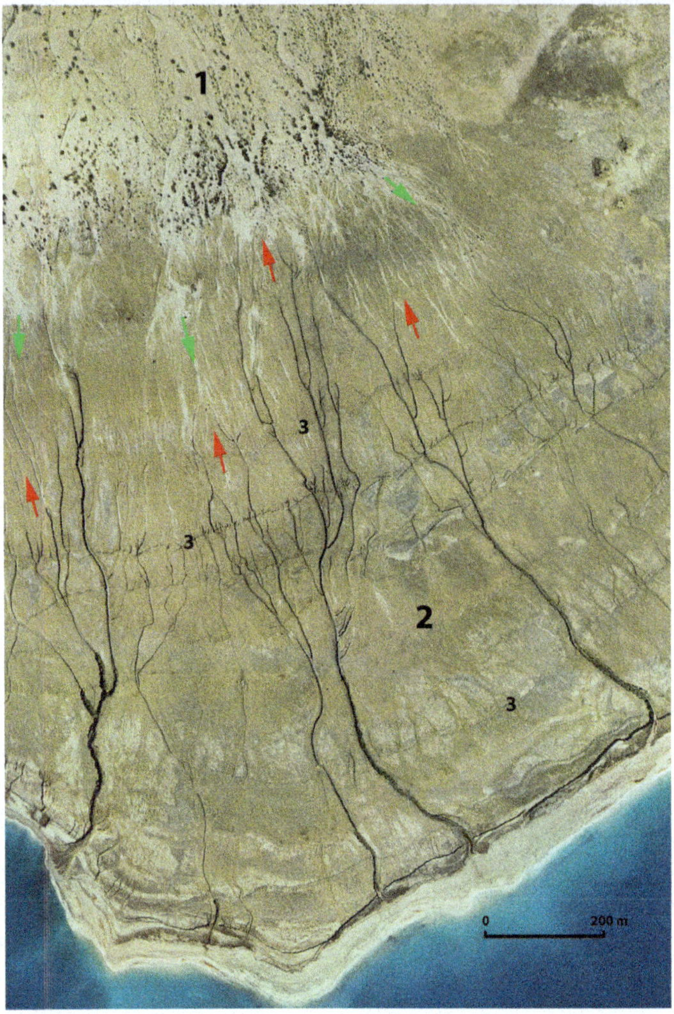

Fig. 16.7 Retreating Dead Sea shoreline in front of Nahal (wadi) Ze'elim (for location see Fig. 16.2): **1** (green arrows)—an active propagating alluvial fan. **2** (red arrows)—the recently exposed lake bed with entrenched channels forming a dendritic pattern triggered by the lowering base-level. **3** Raised shorelines

propagation of the incisional wave, the highway will be destroyed as happened to the bridge on N. Arugot in 2001 (Figs. 16.10 and 16.11). Following the collapse of the bridge, the road was diverted 400 m westward upstream, i.e., an example of a man-made structure *fleeing* in front of an upstream migrating wave of entrenchment.

Fig. 16.8 Aerial view of Nahal Darga outlet (1999) with Highway 90 (for location see Fig. 16.2). The concrete revetment of the highway (**A**) is a local base-level. The alluvial fan surface (**B**) is farmed. The entrenched channel segment downstream from the highway makes a deep trapezoidal cross section with active lateral incision. At the waterline several deep gullies (**C**) cut the stream bed into an island bar (**D**) that merge upstream with the channel bed. Note the incised dirt road

The Dead Sea area demonstrates in addition, the following development (Dente et al. 2017): (a) a hard halite rock is deposited in the lake serving as a local base-level. N. Arava (Fig. 16.2) which enters the Dead Sea from the south becomes graded toward this resistant halite. (b) Later, the halite became breached and stopped to function as a local base-level. (c) The Dead Sea level continued to fall and serve as the active terminal base-level controlling N. Arava.

Flow conditions in the Dead Sea wadis have been recorded by Cohen and Laronne (2005) and by Zituni et al. (2021). Cross-sectional mean flow velocities as high as 6 m s^{-1} have been reported. The following mean vertical entrenchment

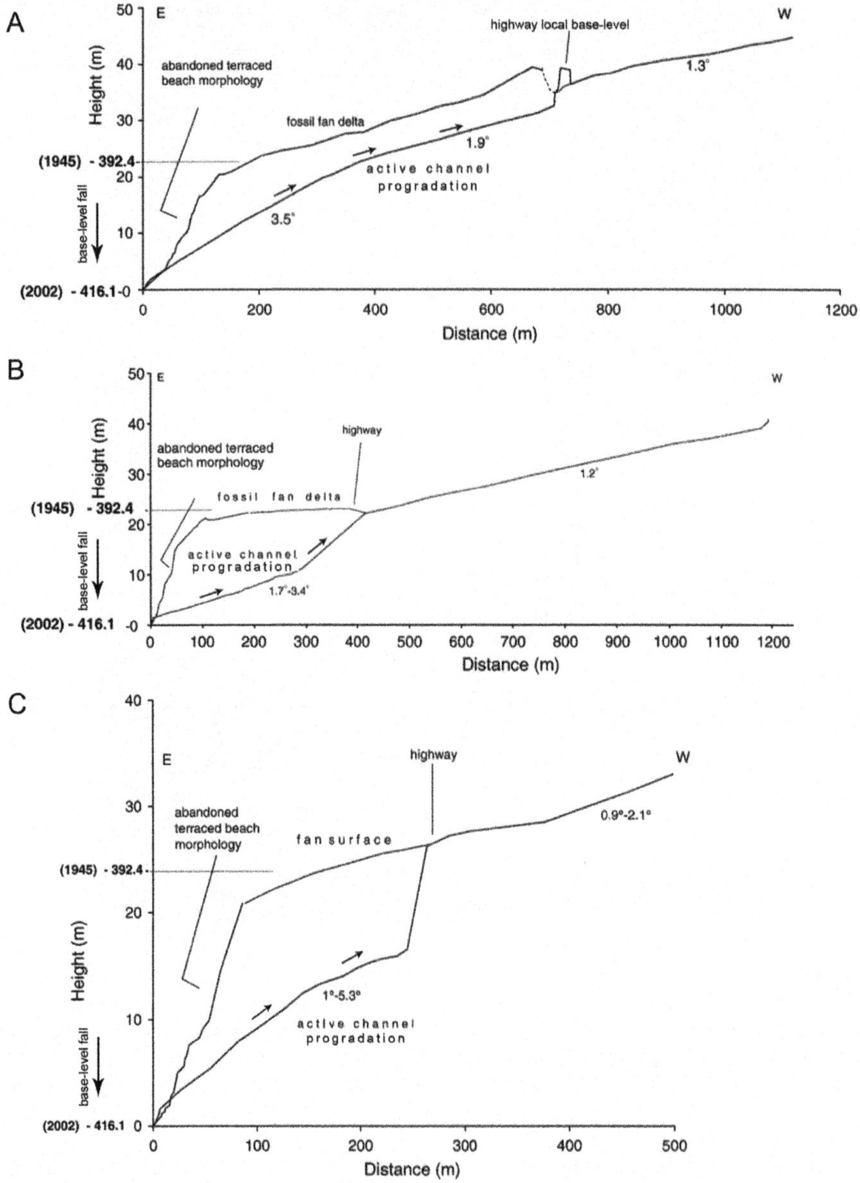

Fig. 16.9 **A**—Nahal Hazezon, **B**—N. Arugot, **C**—N. David: incisional headward propagation following the Dead Sea base-level fall. The highway which serves as a local base-level terminates the headward entrenchment

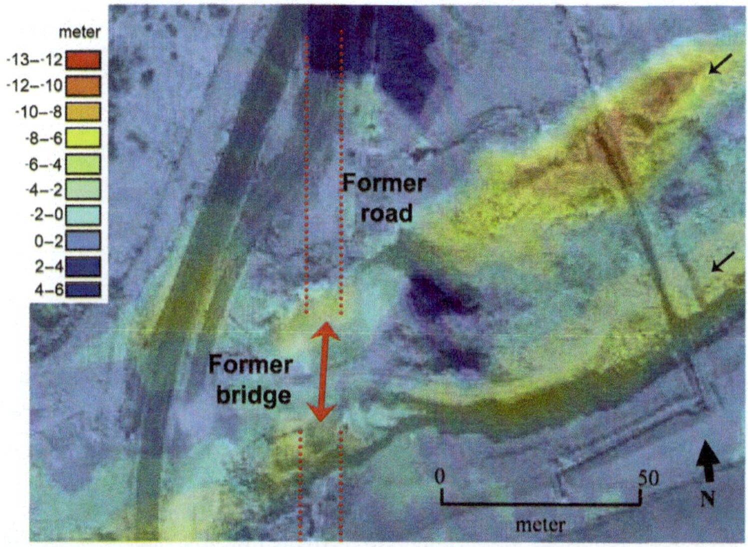

Fig. 16.10 Change detection map of N. Arugot (for location see Fig. 16.2) for the period 1999–2002. The shoreline is located a few hundred meters eastward from the road (not in the map). Note the transient incisional propagating wave along the banks (arrows) that indicate channel widening up to below the bridge that finally collapsed on May 20, 2001, following its detachment from both banks during a large flood (see oblique view at Fig. 16.11). The legend refers to the depth of incision of image cells for the entire three years. Mark the present road (from Bowman et al. 2007)

Fig. 16.11 Oblique view toward the lower section of N. Arugot from above the Dead Sea, following the May 2001 flood. The site of the collapsed bridge is encircled. A trapezoidal shape controls the channel. Cross sections 4, 5 are about 90 m wide

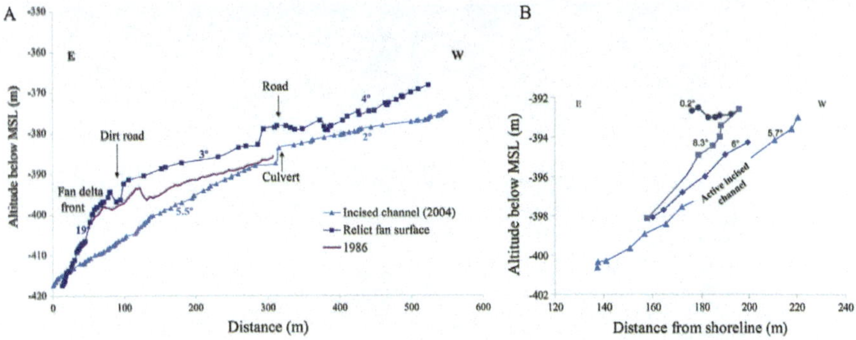

Fig. 16.12 A—Longitudinal profiles of channel beds of the N. Qedem fan delta, including the DEM-based reconstructed 1986 channel bed, the 2004 incised channel bed and the uppermost relict surface demonstrating the high steepness of the fan delta environment. **B** Profiles of channel terraces at the fan delta front. Total incision at this site is 10 m (from Storz-Peretz et al. 2010)

rates have been reported: 0.5–0.7 m/year vertical incision rate in N. Arugot and N. Ze'elim (Bowman 2010); incision rates of 0.3–0.5 m/year based on field cross section in N. Hazezon, N. David and N. Temarim (Devora 2006; Fig. 16.2). A different, high incision rate was recorded at N. Qedem, near the Dead Sea coast, based on change detection maps and channel cross sections based on DEMs. Here the vertical incision rates were as high as 0.7–0·8 m $year^{-1}$ while the lake-level fall rate occasionally rose above 1 m $year^{-1}$. The highest channel vertical incisional rates occurred near the shore, approximating the rate of the base-level drop.

The Qedem (Fig. 16.2) illustrates an extreme incisional example in global scale. Maximum reported rates of incision in alluvium and in silt beds range 20–50 mm $year^{-1}$ (Phillips et al. 2005; Simon and Rinaldi 2006). The incision rates in N. Qedem are much higher in spite of its coarse-textured bed and small (12 km^2), arid drainage basin, expressed by its low rain index of $RI = 1{\cdot}4 \times 106\ m^3\ year^{-1}$, which is the mean annual precipitation volume over the basin (Ben-Moshe 2005). The vertical incision rates and the total entrenchment of N. Qedem are similar to those of much larger basins such as N. Arugot and N. Ze'elim, with respective drainage areas of 232 and 260 km^2 and of much higher rain indices. The incision rate in the Qedem seem not to be controlled by its drainage area as the proxy for sediment and water discharge. It is suggested that the high gradient of the Qedem fan-delta (Fig. 16.12) related to the clinoform structure leads to these extremely high incision rates and demonstrates the importance of the initial morphology and the structure when responding to base-level lowering.

16.4 Longitudinal Profiles and Connectivity

The length of the headward entrenched reaches, activated by the Dead Sea drop, was in N. Hever and N. Ze'elim several hundreds to 1000 m (Fig. 16.13). Agreeing with previous studies (Leopold and Bull 1979) that reported, under base-level drop of tens of meters, entrenchment to short distances upstream.

The bathymetry which became exposed when the Dead Sea dropped, determined the incisional response. The common geomorphic assumption is that a rapid base-level fall across a steep bathymetry will intensify incision and result in divergence of profiles in the downflow direction. The opposite, i.e., convergence of profiles downstream, implies an improved adjustment of the channel to the base-level associated with deceleration in the rate of base-level fall, or with its stability (Fig. 9.4). Terrace remnants were surveyed in the Dead Sea area on rectified aerial photographs by Ben-Moshe et al. (2008) who, for terrace correlation used optically stimulated luminescence (OSL) ages and soil stratigraphy. He showed divergence of profiles (Fig. 16.14) reflecting the rapid, constantly dropping Dead Sea base-level. Hassan and Klein (2002) studied the response of the lower Jordan River to the fall of the Dead Sea level and reported too diverging profiles. Dente et al. (2018) demonstrated profiles of the Jordan river becoming convex

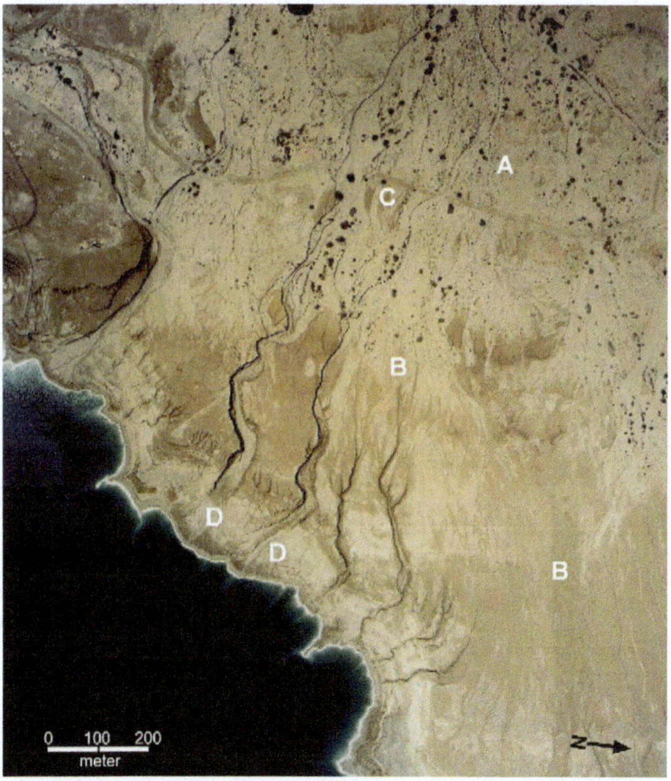

Fig. 16.13 Front of the Ze'elim alluvial fan (**A**) prograding toward the rapidly (> 1 m/year) receding Dead Sea (2006). Short channels are entrenched in the exposed lake bed (**B**), pinching out upstream (**C**) and depositing small alluvial fans (**D**) toward the falling Dead Sea level

and diverging with time (Fig. 16.15). Ben-Moshe and Lensky (2022) reported a convex-diverging pattern of the longitudinal profiles of N. Kidron (south to N. Kumran). Fitting the longitudinal channel profiles of Nahal Darga, N. Kedem and N. Hever to linear and logarithmic curves showed that close to their outlet, the profiles are linear to slightly convex (Bowman et al. 2007).

In profile analysis, deviation from concavity is usually a reliable indicator of extrinsic controlling factors such as tectonics, resistant lithology, or coarse bed-load deposited by tributaries (Seidl et al. 1994). Channel convexities in the Dead Sea area showed no discernable correlation to neotectonic lines nor to lithological control suggesting that convexity results in many cases from the continuous rapid

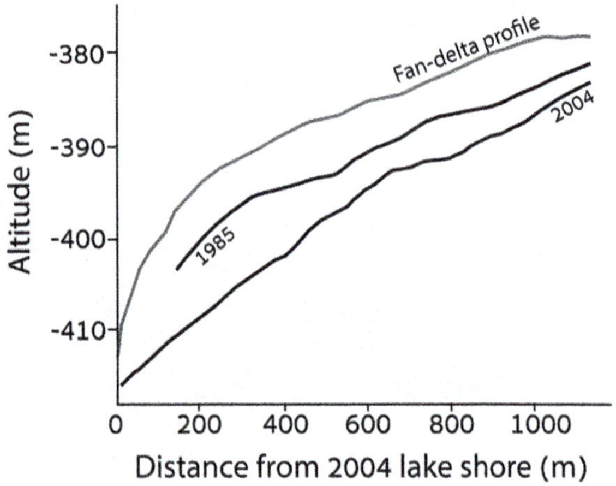

Fig. 16.14 Convex 1985 and 2004 longitudinal profiles of Nahal Qidron entrenched in its fan Delta (redrawn from Ben-Moshe et al. 2008)

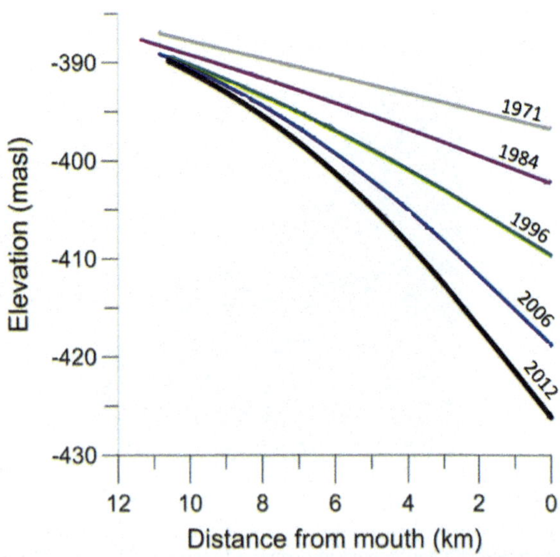

Fig. 16.15 Longitudinal thalweg profile evolution of the Jordan river entering the Dead Sea from the north: the profiles become convex and diverge with time. Gradients range 0.001–0.005. Not shown is the increasing rate of the Dead Sea level fall accompanied by exposure of steeper parts of the delta front (from Dente et al. 2018)

drop of the Dead Sea level. Convexity (discussed in Chap. 9.1) may be as well controlled by the structural convex front of steep, erodible, small and coarse fan-deltas exposed and entrenched along the Dead Sea shore (Storz-Peretz et al. 2011; Ben-Moshe et al. 2008; Bowman et al. 2007).

Connectivity of channels in arid areas to a rapid, continuous base-level fall is dependent on the rainfall regime. Small drainage basins may be episodically omitted by spotty rain cells and provide high probability for dry years. The gap between the recurrence intervals of floods and the continuous rapid base-level drop may cause disconnectivity between base-levels and channels. The connectivity of channels to the ongoing, rapid base-level drop was studied on the steep-fronted, erodible, small and coarse fan-delta of Nahal Qedem, nourished by a hyper-arid, small watershed (12 km^2) with an extreme low rain index, RI = 1.4 × 10^6 m^3 $year^{-1}$ (Storz-Peretz et al. 2010). The coastline of N. Qedem demonstrated a *hanging valley* indicating disconnectivity between the channel outlet and the shoreline (Fig. 16.16). The channel recovered from its *elevated* position and equilibrated to the receding base-level within the next single flow event, demonstrating that the *hanging position* does not limit transmission of the base-level fall effect headward over a period longer than a few dry years.

Another example of disconnectivity is shown by N. Kumran with its small arid to hyper-arid climate (Eyal et al. 2019). Because of its small capacity for incision and sediment transport, N. Kumran does not keep pace with the regressing shoreline and its channel is disconnected from the rapid receding lake. Another example is offered by alluvial fan terraces. The largest and best preserved sequence of 14 fan terraces was morphologically studied in N. Ze'elim (Bowman 1988). In

Fig. 16.16 View southward along the Dead Sea shoreline at the Qedem outlet: the episodically elevated Qedem channel is "hanging" above, detached from the dropping lake level

the Dead Sea area, the longitudinal stream profiles have been generally related to entrenchment triggered by the lowering lake (Willis 1928; Nir 1965; Sneh 1979). There is, however, evidence that the fan terraces are part of the *inherent fluviatile fan system* and should not be attributed to base-level drops. This is based on the significant difference between the number of the Lisan-raised beaches 40 (Bowman 1971) and that of the fan terraces in the largest sequences (18 terraces in N. Hemar, 14 in N. Ze'elim and 12 in N. Lot). The difference may suggest that we are dealing with two unrelated systems and that the fan progradation lagged behind the rapid lake level drop at its front (Bowman 1988). Such evolution fits the non-incisional base-level fall case, postulated by Harvey (2002), in which base-level fall occurs in front of an arid, seldom active, alluvial fan, that does not keep pace with the rapid falling base-level.

16.5 Sinuosity

Demonstrated by laboratory results (Shepherd and Schumm 1974; Ouchi 1985) as well as by field observations (Burnett and Schumm 1983), a steepened channel will increase its sinuosity by lateral erosion in order to restore its original flow gradient. Later, Schumm (1993) suggested his conceptual model of sinuosity evolution, accordingly sinuosity is expected to evolve to 1.5–2.0. Along the western Dead Sea shore, no enhanced thalweg sinuosity beyond 1.1–1.2 had been recorded (Bowman et al. 2007). The sinuosity of the newly formed ~10-km-long downstream segment of Nahal Arava is also not beyond 1.2 (Dente et al. 2017). Increased sinuosity has been shown by Hassan and Klein (2002) in the newly exposed steep Jordan delta. They reported sinuosity increase from ~1.05 to 1.35 during the period 1960–1995.

Following the Dead Sea fall ,emergence of different bathymetric slopes provided study areas to directly examine the sinuosity response to steepening. In the Jordan River, Dente et al. (2018) showed a sinuosity increase near the mouth of the river, at a ~30 m channel depth, where overbank flow was prevented and lateral erosion along the weak, highly erodible channel bank walls became significant. In the upstream segment, where high-magnitude floods still reach the floodplain, the channel straightened out through cutoffs. Channels fed by perennial springs, entrenched in the erodible substrate exposed following emergence of preexisting lake bathymetry, enabled further study of the impact of diverse gradients on channel pattern evolution (Dente et al. 2021). The highest sinuosity evolved in the steepest and deepest channels under confined conditions and scarce cutoffs.

16.6 Cross-Sectional Evolution

What is the cross-sectional response of the channels to the entrenchment forced by the rapid Dead Sea level fall? Previous experiments showed that a rapid base-level drop enhances vertical incision relative to lateral activity (Yoxall 1969). The maximum vertical incision is expected closest to the lowering base-level (Leopold and

Bull 1979; Begin et al. 1981). A shift from vertical to lateral entrenchment would reduce the hydraulic depth and increase energy dissipation (Simon and Darby 1997).

Cross sections in N. Qedem, close to the shore are V-shaped and widen upstream showing a *wine-glass pattern* (Beaty 1961; Fig. 16.17). Such pattern fits the stratigraphic structure of N. Qedem fan-delta, i.e., wider cross sections on the smoother topsets narrow at the front along the steep plunging clinoform structure at the prograding front of fan deltas. The cross-sectional evolution in N. Arugot (Fig. 16.18) demonstrates a gradual change toward a trapezoidal cross section. In N. Ze'elim, the trapezoidal form dominates as well (Fig. 16.19; Shapira 2007). Widening toward a trapezoidal cross-sectional shape has been also observed by change detection maps of N. Qedem (Fig. 16.20). Widening was also observed by Ben-Moshe et al. (2008). However, increase in width did not result in a significantly larger wetted perimeter, i.e., no decreased hydraulic depth, nor a significant increased roughness. In N. Og (Fig. 16.20) and in the Jordan river the picture is different: here mainly vertical incision dominates. Dente et al. (2018) reported at the downstream ~ 2 km reach of the Jordan River, a vertical incisional trend in the last 45 years. The cross section changed from a shallow channel into a ~ 30 m-deep gorge and narrowed from ~ 40 to ~ 15 m during this period.

Channel widening, shown by the trapezoidal cross sections, demonstrated as well steep banks, tension cracks, failed slabs and frequent bank failures (Shapira 2007). These result from the cohesionless sandy-gravelly deposits that make up the erodible substrate in the Dead Sea shore and is unable to sustain banks to confine the flow. *Trapezoidal widening* in the Dead Sea Area does not indicate the stage of transition from deepening to widening, i.e., it does not mark the final evolutionary phase conceptualized by models. Further widening is retarded by the continuous, rapid base-level drop. The channels are entrapped in deepening, prevented from reaching the final wide, graded, stable cross-sectional geometry (Bowman et al. 2010). Under such conditions, previous conceptual models do not apply in full.

16.7 Summary

The Dead Sea area with its dramatic, recent, man-made base-level fall provides unstable conditions which do not allow reaching steady state. The lake provides excellent examples of drainage networks that respond continuously to a rapid and accelerating base-level fall and remain continuously trapped in disequilibrium.

The hundreds of meters elevation-difference between the shoulders of the rift and the Dead Sea level do often not provide special power or erosional efficiency. Many *local base-levels* along the wadis draining to the Dead Sea are those controlling the drainage networks, part of them are waterfalls such as in N. Kedem 340 m high; in N. David 180 m; in N. Mishmar 160 m and in N. Hever 130 m high (Haviv et al. 2006). In N. Zin a 20-m-high vertical waterfall carved into hard limestone bounds the upper Zin drainage basin whereas the southern Dead Sea basin acts as a second local base-level (Davis 2009; Fig. 16.21). The Dead Sea

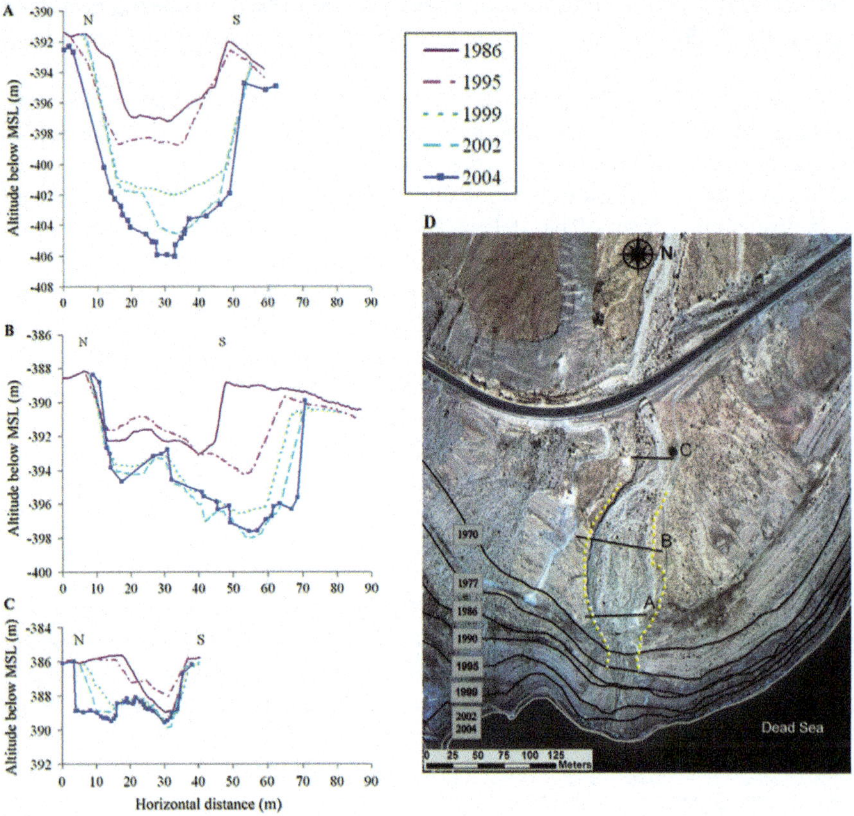

Fig. 16.17 N. Qedem—cross sections based on field survey (2004) and on DEM reconstructions (1986–2004). **A** v-shaped cross section indicating vertical entrenchment with minor widening. Cross sections **B**, **C** demonstrate widening towards a trapezoidal form. The most advanced stage of channel evolution, controlled by widening and deposition, indicating proximity to steady state, has not been observed yet. Demonstrated is a "wine-glass" channel pattern (yellow) that follows the topsets (wide) and the steep clinoform front (narrow) of the fan delta (from Storz-Peretz et al. 2011)

area thus exhibits the efficiency and *importance of local over terminal base-levels.* The highest local base-level is the one that controls most of the network. Being located at the lowest site on earth, hundreds of meters below the global mean sea level, is for the Dead Sea often hydraulically meaningless as a terminal base-level.

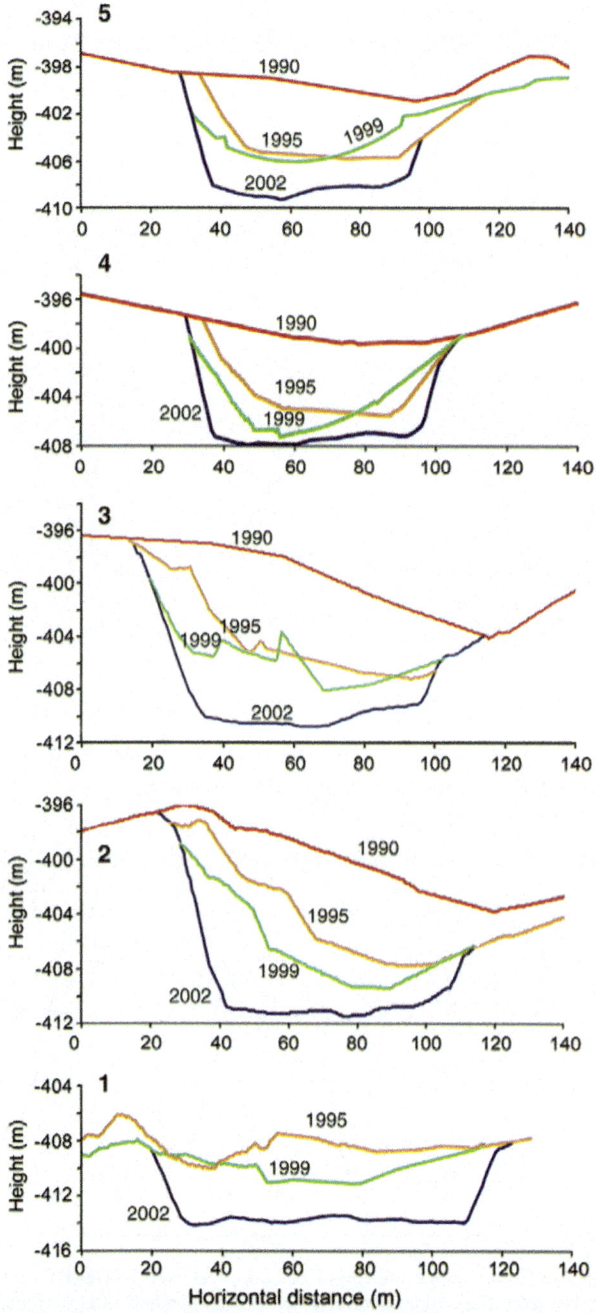

Fig. 16.18 DEM-based cross-sectional evolution along the lower reach of N. Arugot. The incision and evolution toward a trapezoidal shape is demonstrated in different time intervals. The sites of the cross sections are demonstrated in Fig. 16.11

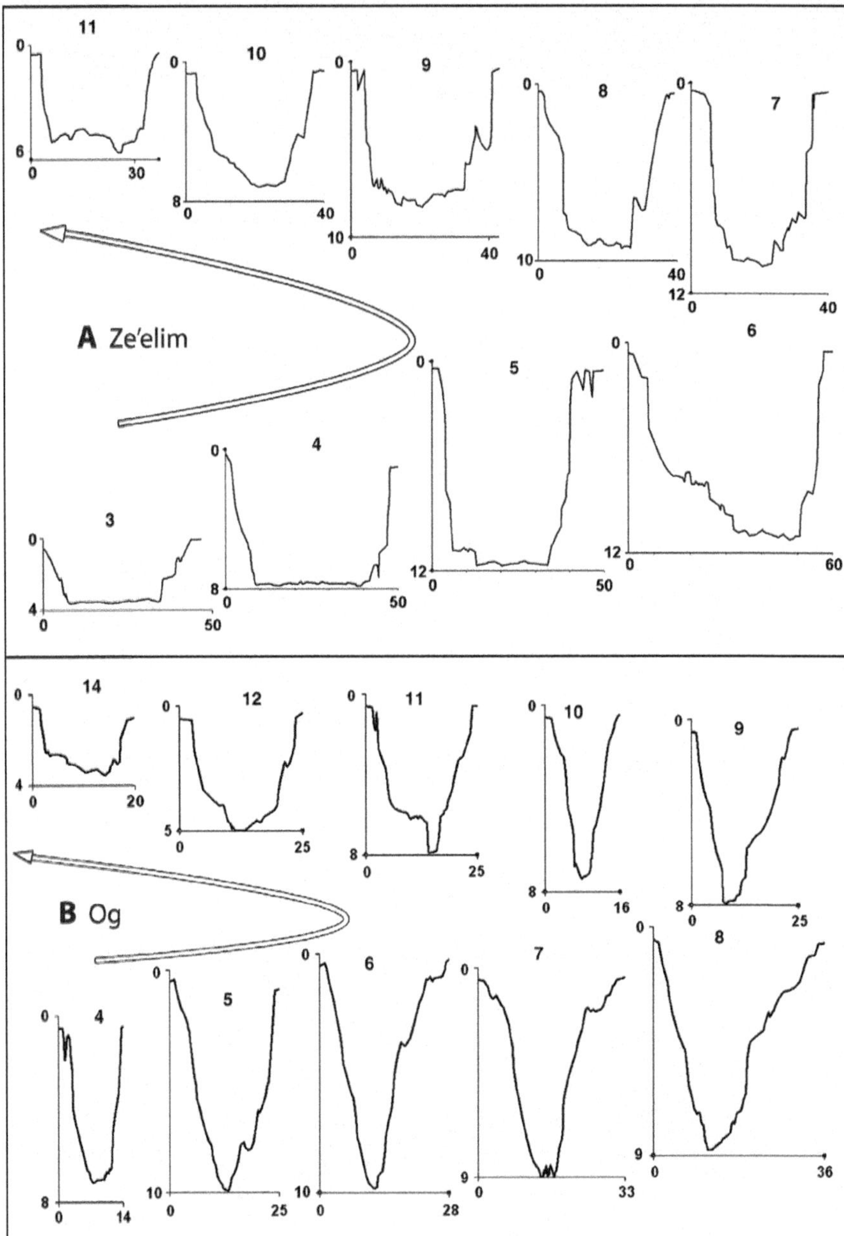

Fig. 16.19 Cross-sectional shapes in meter of N. Zeelim (**A**) and N. Og (**B**). Numbering is headward along 1–2 km from the outlet. Both channels are entrenched in lacustrine, silty-loam lake beds. V-shaped cross sections dominate N. Og, different from the more trapezoidal N. Ze'elim sections. Based on field surveys. For location see Fig. 16.2 (following Shapira 2007)

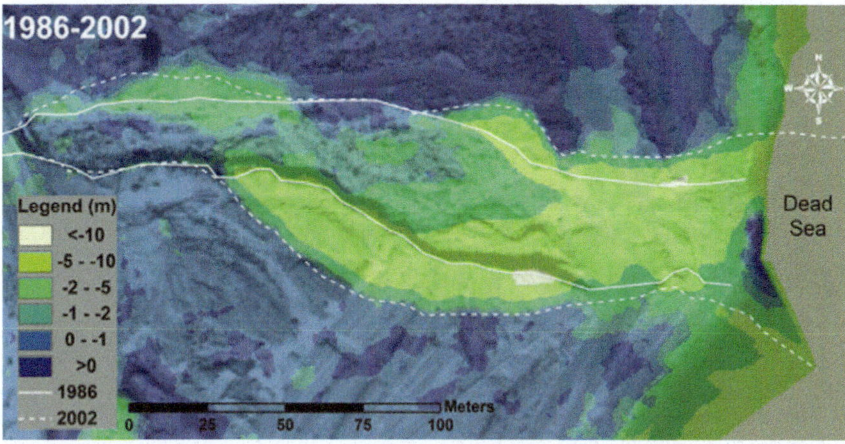

Fig. 16.20 Headward incision from the coastline of the Dead Sea westward into the channel bed of N. Qedem shown by a change detection map. The white dashed lines indicate the banks at 2002. The map shows the progradation of the headward incision along the banks. The legend refers to the depth of incision for the entire 16 years (from Storz-Peretz et al. 2010)

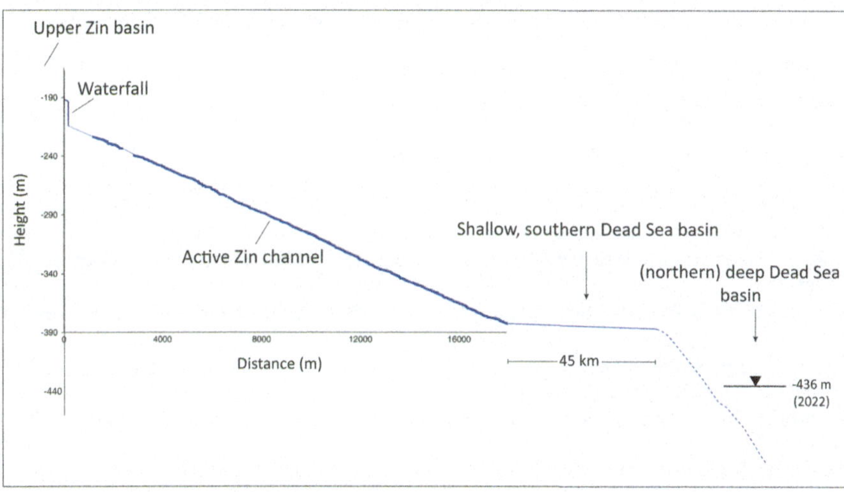

Fig. 16.21 Longitudinal profile of the Zin channel controlled by two local base-levels and uncoupled to the Dead Sea terminal level. The Dead Sea basins are not to scale (redrawn from Davis et al. 2009)

References

Ashbel D (1965) The rise and fall of the Dead Sea lake level. Mada 9:255–260 (in Hebrew)

Bar O, Zilberman E, Feinstein S, Calvo R, Gvirtzman Z (2016) The uplift history of the Arabian Plateau as inferred from geomorphologic analysis of its northwestern edge. Tectonophysics 671: 9–23

Bartov Y, Stein M, Enzel Y, Agnon A, Reches Z (2002) Lake levels and sequence stratigraphy of Lake Lisan, the late Pleistocene precursor of the Dead Sea. Quatern Res 57(1):9–21

Beaty CB (1961) Topographic effects of faulting, Death Valley, California. Ann Assoc Am Geogr 51:234–240

Begin ZB, Meyer DF, Schumm SA (1981) Development of longitudinal profiles of alluvial channels in response to base-level lowering. Earth Surf Proc Land 6:49–68

Ben-Moshe L (2005) Longitudinal profiles of alluvial streams in response to twentieth century changes in the Dead Sea. Unpublished master thesis, Hebrew University of Jerusalem, Israel (in Hebrew)

Ben-Moshe L, Haviv I, Enzel Y, Zilberman E, Matmon A (2008) Incision of alluvial channels in response to a continuous base-level fall: field characterization, modeling and validation along the Dead Sea. Geomorphology 93:524–536

Ben-Moshe L, Lensky N (2022) Geomorphological principles for artificial adjustments to alluvial channels slope. Examples from Dead Sea tributaries. Ministry of Energy, Geological Survey of Israel, GSI-12-2022

Bowman D (1971) Geomorphology of the shore terraces of the late Pleistocene Lisan lake (Israel). Palaeogeogr Palaeoclimatol Palaeoecol 9:183–209

Bowman D (1977) Stepped bed morphology in arid gravelly channels. Bull Geol Soc Am 88:291–298

Bowman D (1978) Determination of intersection points within a telescopic alluvial fan complex. Earth Surf Process 3:265–276

Bowman D (1988) The declining but non-rejuvenating base-level—the Lisan lake, the Dead Sea area, Israel. Earth Surf Process Land 13(3):239–249

Bowman D (1990) Climatically triggered gilbert-type lacustrine fan deltas, the Dead Sea area, Israel. Special Publications. International Association of Sedimentologists, vol 10, pp 273–280

Bowman D (1997) Geomorphology of the Dead Sea western margin. In: Niemi TM, Ben Avraham Z, Gat J (eds) The Dead Sea: the lake and its setting. Oxford University Press, Oxford, pp 217–225

Bowman D (2010) The Dead Sea graben: geomorphology of the lowest spot on Earth. In: Migon P (ed) Geomorphological landscapes of the world. Springer, Berlin, pp 247–255

Bowman D, Devora Sh, Svoray T (2011) Drainage organization on the newly emerged Dead Sea bed, Israel. Quatern Int 233:53–60

Bowman D, Gross T (1992) The highest stand of lake Lisan: ~150 meter below MSL, Israel. J Earth Sci 41:233–237

Bowman D, Shachnovich-Firtel Y, Devora S (2007) Stream channel convexity induced by continuous base-level lowering, the Dead Sea, Israel. Geomorphology 92(1–2):60–75

Bowman D, Svoray T, Devora S, Shapira I, Laronne JB (2010) Extreme rates of channel incision and shape evolution in response to a continuous, rapid base-level fall, the Dead Sea, Isreal. Geomorphology 114(3):227–237

Briaud JL, Ting F, Chen HC, Cao Y, Han SW, Kwak K (2001) Erosion function apparatus for scour rate predictions. J Geotech Geoenviron Eng 127(2):105–111

Burnett AW, Schumm SA (1983) Alluvial-river response to neotectonic deformation in Louisiana and Mississippi. Science 222(4619):49–50

Cohen H, Laronne JB (2005) High rates of sediment transport by flashfloods in Southern Judean Desert, Israel. Hydrol Process 19:1687–1702

Davis M, Matmon A, Zilberman E, Porat N, Gluck D, Enzel Y (2009) Bathymetry of Lake Lisan controls late Pleistocene and Holocene stream incision in response to base level fall. Geomorphology 106(3–4):352–362

Dente E, Lensky NG, Morin E, Grodek T, Sheffer NA, Enzel Y (2017) Geomorphic response of a low-gradient channel to modern, progressive base-level lowering: Nahal HaArava, the Dead Sea. J Geophys Res Earth Surf 122(12):2468–2487

Dente E, Lensky NG, Morin E, Dunne T, Enzel Y (2018) Sinuosity evolution along an incising channel: new insights from the Jordan River response to the Dead Sea level fall. Earth Surf Proc Land 44(3):781–795

Dente E, Lensky NG, Morin E, Enzel Y (2021) From straight to deeply incised meandering channels: slope impact on sinuosity of confined streams. Earth Surf Proc Land 46(5):1041–1054

Devora Sh (2006) The efficiency of the Dead Sea as a lowering base-level. Unpublished master thesis, Ben-Gurion University of the Negev, Beer Sheva, Israel (in Hebrew)

Elias E, Gavrieli I, Lensky NG, Abelson M, Ganor J, Oren A, Yechieli Y (2011) Red Sea to Dead Sea water conveyance (RSDSC) study: Dead Sea research team: geological survey of Israel and Tahal Group. Geol Surv Isr Rep 163:372–437

Eyal H, Dente E, Haviv I, Enzel Y, Dunne T, Lensky NG (2019) Fluvial incision and coarse gravel redistribution across the modern Dead Sea shelf as a result of base-level fall. Earth Surf Proc Land 44(11):2170–2185

Garfunkel Z (2014) Lateral motion and deformation along the Dead Sea transform. In: Garfunkel Z, Ben-Avraham Z, Kagan E (eds) Dead Sea transform fault system. Springer, Berlin, pp 109–150

Ghazleh SA, Kempe S (2009) Geomorphology of Lake Lisan terraces along the eastern coast of the Dead Sea, Jordan. Geomorphology 108(3–4):246–263

Greenbaum N, Ben-Zvi A, Haviv I, Enzel Y (2006) The hydrology and paleohydrology of the Dead Sea tributaries. In: Enzel Y, Agnon A, Stein M (eds) New frontiers in Dead Sea paleoenvironmental research. Geological Society of America, Special paper 401, pp 63–93

Haase-Schramm A, Goldstein SL, Stein M (2004) U-Th dating of Lake Lisan (late Pleistocene Dead Sea) aragonite and implications for glacial East Mediterranean climate change. Geochim Cosmochim Acta 68(5):985–1005

Harvey AM (2002) The role of base-level change in the dissection of alluvial fans: case studies from southeast Spain and Nevada. Geomorphology 45(1–2):67–87

Hassan MA, Klein M (2002) Fluvial adjustment of the lower Jordan River to a drop in the Dead Sea level. Geomorphology 45(1–2):21–33

Haviv I, Enzel Y, Whipple KX, Zilberman E, Stone J, Matmon A, Fifield LK (2006) Amplified erosion above waterfalls and oversteepened bedrock reaches. J Geophys Res Earth Surf 111(F4)

Kaufman A (1971) U-series dating of Dead Sea basin carbonates. Geochim Cosmochim Acta 35(12):1269–1281

Klein C (1986) Fluctuations of the level of the Dead Sea and climatic fluctuations during historical times. PhD thesis, Hebrew University of Jerusalem, Israel (in Hebrew)

Klinger Y, Avouac JP, Bourles D, Tisnerat N (2003) Alluvial deposition and lake-level fluctuations forced by Late Quaternary climate change: the Dead Sea case example. Sed Geol 162(1–2):119–139

Lensky N, Dente E (2015) The hydrological processes driving the accelerated Dead Sea level decline in the past decades. Geological Survey of Israel Report

Leopold LB, Bull WB (1979) Base-level, aggradation and grade. Proc Am Philos Sci 123:168–202

Masterman EWG (1913) Summary of observations on the rise and fall of the level of the Dead Sea 1900–1913. Palestine Explor Q 192–197

Matmon A, Crouvi O, Enzel Y, Bierman P, Larsen J, Porat N, Caffee M (2003) Complex exposure histories of chert clasts in the late Pleistocene shorelines of Lake Lisan, southern Israel. Earth Surf Process Land J Br Geomorphol Res Group 28(5):493–506

Neev D (1964) Recent sedimentary processes in the Dead Sea. Thesis submitted for the Degree Doctor of Philosophy, Hebrew University

Neev D, Emery KO (1967) The Dead Sea. Geol Surv Israel Bull 41:1–147

Nir D (1965) A geomorphological map of the Judean desert 1:100,000. Scripta Hierosolimitana IS: 5-29

Ouchi S (1985) Response of alluvial rivers to slow active tectonic movement. Geol Soc Am Bull 96(4):504–515

Phillips JD, Slattery MC, Musselman ZA (2005) Channel adjustment of the lower trinity River, Texas, downstream of Livingstone Dam. Earth Surf Proc Land 30:1419–1439

Quennell AM (1958) Tectonics of the Dead Sea Rift. Congresso Geologico Internacional, Ciudad De Mexico, 1956. Association Services Geologiques Africains, pp 386–405

Schramm A, Stein M, Goldstein SL (2000) Calibration of the 14C time scale to> 40 ka by 234U–230Th dating of Lake Lisan sediments (last glacial Dead Sea). Earth Planet Sci Lett 175(1–2): 27–40

Schumm SA (1993) River response to base-level change: implications for sequence stratigraphy. J Geol 101:279–294

Schumm SA, Chorley RJ (1983) Geomorphic controls on the management of nuclear waste. U.S. Nuclear Regulatory Commission, Washington, D.C. NUREG, p 137

Seidl MA, Dietrich WE, Kirchner JW (1994) Longitudinal profile development into bedrock: an analysis of Hawaiian channels. J Geol 102:457–474

Shapira I (2007) Cross sectional shape and bank processes in micro canyons incised into lacustrine sediments in the north western Dead Sea shore. Unpublished MA thesis, Ben-Gurion University of the Negev, Beer Sheva, Israel (in Hebrew)

Shepherd R, Schumm SA (1974) Experimental study of river incision. Geol Soc Am Bull 85(2):257–268

Simon A, Darby SE (1997) Process-form interactions in unstable sand-bed river channels: a numerical modeling approach. Geomorphology 21(2):85–106

Simon A, Rinaldi M (2006) Disturbance, stream incision, and channel evolution: the roles of excess transport capacity and boundary materials in controlling channel response. Geomorphology 79:361–383

Sneh A (1979) Late Pleistocene fan deltas along the Dead Sea Rift. J Sediment Petrol 49:541–552

Storz-Peretz Y, Bowman D, Laronne JB, Svoray T (2010) Rapid incision of a small, coarse and steep fan-delta in response to base-level fall: the case of Nahal Qedem, the Dead Sea, Israel. Earth Surf Process Land 36(4):467–480

Torfstein A, Goldstein SL, Stein M, Enzel Y (2013) Impacts of abrupt climate changes in the Levant from Last Glacial Dead Sea levels. Quatern Sci Rev 69:1–7

Willis B (1928) Dead Sea problems—rift valley or ramp valley. Bull Geol Soc Am 39:49C – 542

Yoxall WH (1969) The relationship between falling base-level and lateral erosion in experimental streams. Geol Soc Am Bull 80(7):1379–1384

Zituni R, Greenbaum N, Porat N, Benito G (2021) Magnitude, frequency and hazard assessment of the largest floods in steep, mountainous bedrock channels of the Southern Judean Desert, Israel. J Hydrol Reg Stud 37:100886

The manufacturer's authorised representative in the EU is Springer Nature Customer Service Centre GmbH, Europaplatz 3, 69115 Heidelberg, Germany. If you have any concerns regarding our products, please contact ProductSafety@springernature.com

Printed and bound by CPI Group (UK) Ltd, Croydon, CR0 4YY
28/09/2026
02224077-0016